Foreword

CIRIA's research programme *Methane and Associated Hazards to Construction* is intended to provide guidance for the construction industry.

In addition to the publication of a bibliography relevant to methane and construction (CIRIA Special Publication 79) and a study of the construction industry's needs for research and information on methane (CIRIA Project Report 5), the programme includes the preparation of two further guidance documents: this report and a companion report on the nature, origins, occurrence and hazards of methane (CIRIA Report 130). Current additional projects deal with the protection of developments from methane and associated gases in the ground and with the investigation of sites where these gases may be present.

This report is the result of the third project in the programme. It was prepared by Dr D Crowhurst and Mr S J Manchester of the Fire Research Station of the Building Research Establishment, under contract to CIRIA.

CIRIA is pleased to acknowledge the contribution to this report arising from longstanding and current work at the Building Research Establishment funded by the Department of the Environment on the potential hazards of landfill gases to construction.

Following CIRIA's usual practice, the research was guided by a Steering Group which comprised:

Mr M G Glynn (Chairman) — North West Water
Mr K V Ansell — McAlpine Laboratories
Mr D L Barry — W S Atkins
Dr J S Edwards — University of Nottingham
Mr D Grant — Department of the Environment
Mr R A B Hall — Mott MacDonald Environmental Services
Mr C P Hill — West Yorkshire Waste Management
Mr A Parker — Gibb Environmental Sciences
Dr J S Sceal — Wardell Armstrong
Mr A A Smith — Clayton Environmental Consultants
Mr R Thompson — London Waste Regulatory Authority
Mr R C Weeks — Geotechnical Instruments

CIRIA's research managers for this project were Mr F M Jardine and Mr R Freer.

The project was funded under Phase I of the programme of *Methane and Associated Hazards to Construction* by:

Department of the Environment, Construction Directorate
National House-Building Council
Anglian Water Services Ltd
Northumbrian Water Services Ltd
North West Water Ltd
Southern Water Services Ltd
Thames Water plc
Welsh Water plc
Yorkshire Water Services Ltd
Kyle Stewart Design Services Ltd
Sir Robert McAlpine & Sons Ltd

CIRIA and Fire Research Station are grateful for help given to this project by the funders, by the members of the Steering Group and by the many individuals and organisations who were consulted.

Acknowledgements

Figure 12 reproduced with the permission of Gordon and Breach Science Publishers.
Figure 19 reproduced with the permission of Supply and Services Canada (1992).

Contents

List of Figures

List of Tables

Glossary

Alkanes	A group of straight chain hydrocarbons, the first four members being the gases: methane, ethane, propane and butane.
Anaerobic	In the absence of oxygen (air).
Bacteriogenic	Derived from ancient or recent microbiological activity on organic matter.
Billion	1×10^9 (1 000 000 000)
Biogenic	Derived from the action of bacteria on organic matter.
Borehole	A hole drilled in or outside the wastes in order to obtain samples. Also used as a means of venting or withdrawing gas.
Catalyst	A substance which speeds up a chemical reaction without itself undergoing any permanent change.
Coulometer	An instrument for measuring the amount of charge passing in an electrical circuit.
Cover	Material used to cover solid wastes deposited in landfills.
Electrolyte	A substance which undergoes partial or complete dissociation into ions in solution, and thus acts as a conductor of electricity.
Fissure	A long narrow cleft or crack.
Flammable	Of a substance capable of supporting combustion in air.
Flux	Flow or discharge.
Geochemical	The result of an underground chemical reaction.
Halogenated	The incorporation of halogens (fluorine, chlorine, bromine, iodine) into a chemical species.
Hot spots	Areas of underground combustion and gas generation.
Hydrocarbon	A compound containing carbon and hydrogen only.
Inert	Having only a limited ability to react chemically.
Ionisation	The process of changing a particle with no charge into one with a positve or negative charge, by the removal or addition of electrons.
Isotope	Atoms of the same chemical element which have different atomic weight, i.e. numbers of neutrons but similar chemical properties.
Landfill	The engineered deposit of waste into or on to land. It may eventually provide land which may be used for another purpose.

Leachate	The result of liquid seeping through a landfill and being contaminated by substances in the deposited waste.
Limits of flammability	Concentration range bounded by LEL and UEL (see abbreviations) within which a gas or vapour is flammable at normal atmospheric temperature and pressure.
Luft cell	A gas-filled cell used in some infra-red detectors to select a particular measuring wavelength.
Mercaptans	A group of organic compounds containing sulphur which have strong and unpleasant odours.
Microbial	Small organisms which are only visible under a microscope, such as bacteria, fungi and algae.
Odourant	A substance with the property of effecting the nasal sense of smell.
Oxidation	Reaction of a species with an oxidant − normally, but not necessarily, 'oxygen' from the air.
Paramagnetic	The property of a substance which, when placed in a magnetic field, causes a greater concentration of the lines of magnetic force within itself than in the surrounding magnetic field.
Peristaltic pump	A pump operating by successive wavelike contractions and relaxations.
Permeable	Allowing the passage of a liquid or gas.
pH	A measure of the acidity or alkalinity of a liquid.
Photoionisation	The ejection of an electron from an atom by a quantum of electromagnetic energy.
Pitot tube	Device used to measure speed of flow.
Scintillant	A substance which produces flashes of light when particles collide with it.
Specificity	In the context of instrumentation implies a response to a single or unique component.
Stenching agent	A substance adding a smell to another substance.
Thermogenic	Derived from the temperature degradation of organic matter.
Topography	The study or description of the surface features of a region.

Abbreviations

BASEEFA   British Approvals Service for Electrical Equipment in Flammable Atmospheres.

BDA   British Drilling Association.

BSI   British Standards Institute.

COSHH   Control Of Substances Hazardous to Health.

DIAL   Differential Absorption LIDAR.

IR   Infra-red.

LEL   Lower Explosive Limit − the lower limit of flammability, i.e. the minimum percentage by volume of a mixture of gas in air which will propagate a flame in a confined space, at normal atmospheric temperature and pressure.

LIDAR   Laser Radar.

NICAT   Nickel Catalyst.

OEL   Occupational Exposure Limit − limits related to personal exposure to substances hazardous to health in the air of a workplace.

OS   Ordnance Survey.

UEL   Upper Explosive Limit − the upper limit of flammability, i.e. the maximum percentage by volume of a mixture of gas in air, at normal atmospheric temperature and pressure, which will propagate flame in a confined space.

VOC   Volatile Organic Compounds.

1 Introduction

1.1 BACKGROUND

This report was prepared for CIRIA by the Building Research Establishment. It forms part of work sponsored by CIRIA, industry, and the Department of the Environment which seeks to provide the construction industry with up-to-date guidance about detecting, sampling, measuring and monitoring methane and other gases that may be found in the ground.

This guidance document is one of a series of reports relating to problems of gas in the ground. The reports from other stages of the project are published separately and each document can be read independently as a report in its own right. The documents are also complementary and together provide a state-of-the-art overview of the subject.

1.2 SCOPE

This report provides guidance on the detection, measurement and monitoring of gases in the ground. Although the report is centred on methane, this gas should not be considered in isolation. Other hazardous gases such as carbon dioxide, hydrogen sulphide, carbon monoxide, and hydrogen, may be present with methane or occur separately; the report also provides guidance for these gases.

1.3 HOW TO USE THIS DOCUMENT

The main text of the report, and therefore what is considered to be essential reading for each section is presented in the normal way.

More detailed information, which may be of interest to some readers, is presented in boxes.

Comments on the main text are presented in italics.

As far as possible tables and figures accompany the relevant text.

In most cases it is intended that each section can be read as a whole, independent of other parts of the text. Where necessary, however, essential cross references are given.

A glossary of terms is included at the beginning of this report.

2 Sources of hazardous gases

The problem of hazardous gases in the ground is not limited to landfill sites and particular types of development. Although much attention has been paid to landfill gas in recent years, the potential problems for construction caused by methane in the ground are much wider.

2.1 METHANE

There are a number of different possible sources of methane that could be encountered during a development. The sections below give a brief description of these sources.

2.1.1 Landfill gas

Landfill gas is formed by the decomposition of degradable wastes within landfill sites. Principally a mixture of methane and carbon dioxide, it also includes a large number of minor components.

The landfill gas production process passes through several stages (Figure 1) during which the composition of the gas changes significantly[1]. Initial decomposition depletes the oxygen in the ground, converting it to carbon dioxide. At the start of the second stage all the oxygen has been consumed and the conditions become anaerobic, leading to the production of a mixture of carbon dioxide and hydrogen. It is not until the third stage that methane production starts (methanogenesis). The gas composition during steady-state methanogenesis is about 60% methane and 40% carbon dioxide. Subsequently, in the final stages, gas production declines although it can continue, albeit at a much lower rate, for many years.

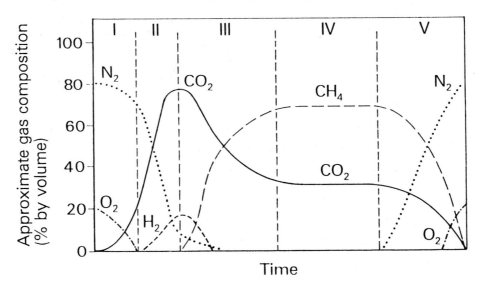

Phase I: Aerobic decomposition of biodegradable materials: entrained atmospheric oxygen is converted to carbon dioxide.
Phase II: Anaerobic decomposition commences as oxygen is used up: carbon dioxide concentration increases and some hydrogen is produced: no methane is produced at this stage.
Phase III: Anaerobic methane production commences and rises to a peak: concentration of carbon dioxide declines: hydrogen production ceases.
Phase IV: Steady methane and carbon dioxide generation in proportions of between 50-70% and 30-50% respectively.
Phase V: Steady decline in generation of methane and carbon dioxide: gradual return to aerobic conditions.

Figure 1 *Phases in landfill gas generation*

Gas production is sensitive to a range of factors such as moisture content, oxygen content, pH, temperature and density of the waste. A more complete description of landfill gas production is given in Waste Management Paper 27[2].

2.1.2 Marine sediments

Methane production from marine sediments occurs due to the anaerobic decomposition of organic matter deposited on the sea bed. Not only must there be no oxygen present for methane to be produced but dissolved sulphate must also be absent. This generally occurs at depths of tens of metres into the sediment in deep sea environments. Methane production at shallow depths is not generally found.

2.1.3 Wetlands

Areas such as marshes, peatlands, lakes and estuaries are all potential sources of methane. The gas formed in these environments predominantly consists of methane and carbon dioxide, sometimes together with hydrogen sulphide. It is formed by the anaerobic microbiological degradation of dead, usually waterlogged, vegetation. Flames of burning marsh gas are responsible for the phenomenon known as Will-o'-the-Wisp.

2.1.4 Sewer gas

The main components of sewer gas are again methane and carbon dioxide, but hydrogen sulphide may also be present. The process of formation is anaerobic decomposition of the organic putrescible components of sewage.

2.1.5 Mine gases

The most common gas encountered in mines and mining areas is methane (firedamp). Geological methane is usually associated with coal-bearing carboniferous strata and is produced by the anaerobic decomposition of ancient vegetation trapped within the rock. Higher alkanes (such as ethane), hydrogen and helium may also be present.

Methane encountered during mining operations can originate from a number of sources[3]:

- as a free gas within the seam being worked on

- adsorbed on the surfaces of fissures and pores in the seam

- absorbed within the internal structure of the coal

- from non-coal strata such as shale or sandstone

- from colliery spoil.

Unworked seams surrounding the seam being mined can also produce methane that will contribute to the hazard in the area being worked.

2.1.6 Groundwater

Under pressure, methane is dissolved in groundwater and is released back into the gaseous phase when the pressure is reduced, rather like the release of gas which occurs on opening a bottle of carbonated water. There is therefore the potential for this gas to escape from solution at the ground surface or in man-made constructions such as tunnels and old mine workings. Methane in groundwater is dealt with in more detail in Section 14.

2.1.7 Natural gas (piped mains gas)

Natural (mains) gas is predominantly methane (with an added stenching agent) which has the same geological source as methane in coal mines. Leaks of this gas from fractured underground mains can be a problem, particularly if the gas subsequently migrates into buildings. The gas pipeline or other utility services often provide preferential routes for this to occur[4]. During the migration of piped gas some of the methane may be oxidised to carbon dioxide.

2.1.8 Other sources

Other potential sources of methane relevant to construction include industrial wastes and certain building materials which contain organic matter. For example foundry sands, timber and paper may produce methane and carbon dioxide if they undergo anaerobic degradation.

2.2 CARBON DIOXIDE

As with methane, carbon dioxide occurring in the ground can come from a number of different sources, as follows:

- landfill — carbon dioxide generally constitutes about 40% of landfill gas and is formed during the initial aerobic decomposition of the refuse, and also later during the anaerobic degradation stage

- wetlands — see Section 2.1.3

- sewers — see Section 2.1.4

- mines — carbon dioxide (blackdamp) can be formed by the low temperature oxidation of carbonaceous materials

- limestone/chalk — carbon dioxide can be produced by the action of acid waters on carbonate rocks as might be found in limestone and chalky areas. Creedy[3,5] gives a fuller discussion of the origins of these gases underground

- other materials containing organic matter — see Section 2.1.8.

2.3 HYDROGEN SULPHIDE

In North West England and the East Midlands hydrogen sulphide groundwater is a fairly common occurrence. In some major urban areas there are natural surface seepages of hydrogen sulphide, often in conjunction with natural methane[6]. The primary sources of hydrogen sulphide are the anaerobic microbial action on sulphur-containing materials such as sulphates and sulphides, and refuse containing gypsum and plasterboard. Where such materials are present in landfills, there is a possibility that hydrogen sulphide may be produced.

2.4 CARBON MONOXIDE

Carbon monoxide can be produced by subterranean fires within combustible wastes such as colliery spoil, peat, and landfill material with a high organic content. Besides the problem presented by the carbon monoxide, the underground fire from which it emanates poses serious problems affecting ground stability. The gas can also be formed by the low-temperature oxidation of carbonaceous material, and hence can be encountered as a component of mine gas.

2.5 OTHER GASES

Hydrogen can be present in landfill gas in small quantities. It is formed with carbon dioxide during the anaerobic process of gas production known as Phase II (see Figure 1).

Landfill gas may also contain trace amounts of saturated and unsaturated hydrocarbons, halogenated compounds, organosulphur compounds and alcohols.

4 Looking for gas – planning an investigation

4.1 INTRODUCTION

The types of site where gas problems may occur are outlined in Section 2. On the assumption that an investigation is warranted, the structure of a good site investigation is considered (Figure 2).

The detailed purpose of a site investigation may vary from site to site, but for most cases it can probably be summarised in the following terms:

1. To confirm and quantify the presence of gas.

2. To establish the requirements for the safe development and utilisation of the area affected by gas.

4.2 SITE INVESTIGATION – THE FIRST STAGE

4.2.1 Desk study

The first and an essential element of any site investigation is a desk study of the site in question. Although this may not need to be extensive it will assist in the efficient planning of further aspects of the gas investigation.

The initial examination should seek to establish the following:

• the history of the site and its use

Some sources of information.

A comprehensive list of sources is given in Reference 8.

Waste Disposal Authorities should be able to provide information on whether the site has been used for landfilling waste. Sets of aerial photographs provide clear indications of excavation and in-filling.

Ordnance Survey maps and plans are available from local and national libraries, with special maps available from Ordnance Survey themselves. Early editions of OS maps can be viewed at the British Library Map Library. County Record Offices may be able to provide other maps and plans.

The local population is often a useful source of information, where landfills are concerned. Residents may have knowledge of 'fly tipping' or other activities not recorded in official documents.

Registers of land which may be contaminated are to be completed and maintained by local authorities under a duty placed on them by Section 143 of the Environmental

---

• the location of underground services

This is an important consideration for the future location of gas sampling points. Care should be taken to avoid underground services when using shallow spiking to probe for gas, and in the construction of permanent sampling points.

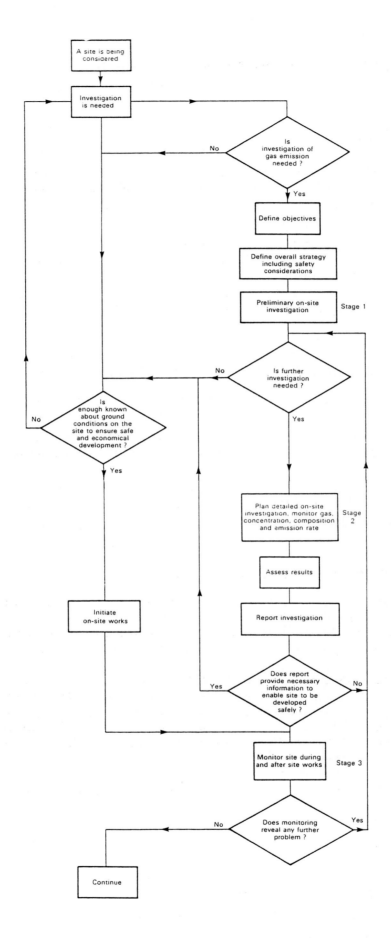

Figure 2 *Sequence of investigation for gas*

- the location of any public rights of way

    Local and national libraries, and county record offices should be able to provide Ordnance Survey maps, plans and charts.

- the types and location of wastes if the site has been filled or is a former landfill

    Important, particularly if the site has been used for the disposal of toxic and chemical wastes.

    The local Waste Disposal Authority should be able to provide details of the amount and type of waste deposited, though they may not be able to if the site was used before 1976 when waste disposal sites were not licensed. Access to the information may in some circumstances be denied on grounds of commercial confidentiality.

    Information on landfill sites operated prior to the introduction of site licensing formerly held by the Regional Water Authorities may now be available from the water companies or the National Rivers Authority.

    The local Environmental Health Department or Planning Department may also be able to supply information.

- the nature of the underlying geology

    Help on the geology and hydrogeology of the site could be obtained from the water companies, British Geological Survey, and the National Rivers Authority.

- if the site is in a mining area, the presence of underground mine workings and the location of disposed spoil

    British Coal (formerly the National Coal Board) should have records of mining activities, although these may not always be complete for an area where private mining has taken place.

Information relating to safety, e.g. the location of underground utilities and information on the presence of toxic materials, should be obtained, as far as is possible, before any on-site measurements are made.

4.3 VISUAL ASSESSMENT

A visit to the site for a visual assessment and to 'sniff' for gas, i.e. take some initial gas readings using a primary investigation instrument such as a flame ionisation detector, is essential. It should be noted, however, that being unable to detect gas by sniffing does not indicate there is not a problem. On old sites there may be no visual evidence of gas activity.

The investigation should not be limited to the site area, but should include testing around its periphery to detect any off-site migration of gases. Any positive detection of gas confirms the need for further investigation. Failure to detect gas at this stage, however, would not necessarily indicate safe conditions for any form of development. For example, there may be a low-permeability layer preventing upward movement of gas, or insufficient ground gas pressure to produce an emission of gas.

In addition, operations on the site might reactivate or modify the gas-producing processes.

Some typical signs of gas might be:

- vegetation die-back
- odours
- bubbling from puddles and ponds
- visible water vapour plumes from cracks in the ground
- areas of melted snow
- signs of burning along cracks

- sulphur deposits

- uneven settlement

- signs of leachate.

Attention should also be paid to the surface topography and layout of the site. The investigator should try to correlate the documentary evidence discovered in the desk study with the existing conditions on the site.

As well as visual evidence it is also worthwhile taking some preliminary gas measurements by spiking with shallow probes and sniffing cracks and fissures for evidence of gas. This can allow quite extensive coverage of the site and may identify potential 'hot spots' of gas emission and generation.

Other techniques which may be of use in obtaining an indication of the general gas condition are described briefly below.

4.3.1 Aerial false-colour infra-red photography

This technique is based on the different infra-red radiation absorption properties of healthy and unhealthy vegetation. On a site where methane is present in the soil the vegetation is often unhealthy, grass sward being particularly affected. The distressed condition of the foliage is not always apparent to the naked eye, but it will show up when photographed using infra-red sensitive film (Figures 3 and 4). To get an accurate picture the site must be photographed from above. Helium-filled balloons[16] and radio-controlled model aeroplanes[17] have been used for this.

Figure 3   *Conventional photograph showing variations in condition of grass (lower left corner). By permission of Skyscan*

It should be stressed that the vegetation could be diseased from causes other than methane, such as leachate from landfill, incorrect use of pesticides and herbicides, and poor irrigation/drainage. As the results obtained from this method are often inconclusive, it is not very widely used. The advantages of this technique are:

- it is relatively inexpensive

- it can be used to examine a wide area

- it can be used to identify migration routes.

Healthy vegetation is a strong infra-red reflector and when photographed will appear red on the film. Distressed foliage records typically as grey-green because of its absorption of the infra-red radiation. As infra-red sensitive film also responds to visible and ultra-violet light, it is necessary to use the appropriate filters so that only images from the near infra-red (500 to 900 nanometres) part of the spectrum are formed.

Figure 4   *False-colour infra-red photograph showing distress in trees and grass (seen in red). By permission of Skyscan*

4.3.2 Aerial thermography

A similar technique to aerial infra-red photography is aerial thermography[16] which involves the use of an infra-red scanner instead of a conventional camera (Figure 5, and Figure 6 at the end of Section 4).

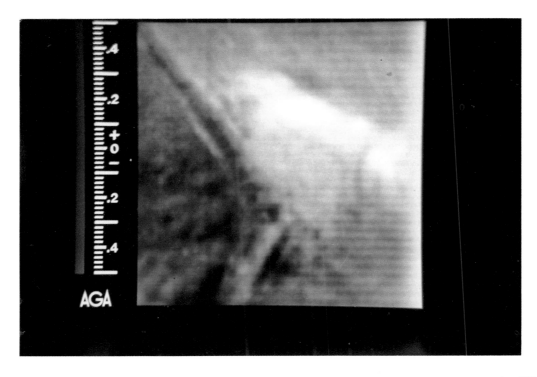

Figure 5   *Infra-red thermogram showing methane emission adjacent to a track on a landfill site. By permission of Wimpey Environmental*

This technique can be used to locate areas of underground combustion and methane generation i.e. 'hot-spots', as these areas will have higher surface temperatures than non-gas-producing areas.

The major disadvantage with this method is that it has to be carried out at night as the sun will mask any hot-spots during the day.

4.4 SECOND-STAGE INVESTIGATION

If the presence of gas on a site is confirmed by a preliminary investigation or where a site is being considered for any form of hard development, further more detailed investigation will be required to quantify the gas problem.

By the end of this part of the examination the investigator should aim to have a clear idea of the gas regime and be able to advise a developer of the potential hazards from any sources of gas and give some indication of the degree of protection that is likely to be required for the proposed development.

The following basic measurements and supporting information are required at this stage:

- gas composition

- gas concentration

- gas emission rates

- the likely variation in the above as a result of normal meteorological changes

- variations with time

- identification of the source of gas

- the extent of any lateral migration

- the presence of methane or other gases in the groundwater

- the nature of the waste

- the geology/hydrogeology of the site.

In order to achieve this, the investigator will have to employ a variety of techniques and types of instrumentation to build up a picture of the overall site conditions.

The instrumentation and techniques for constructing sampling points and sampling methods are discussed in detail in Sections 7 to 12.

The choice of which technique to use will be governed by the nature of the information needed and the requirement to protect the health and safety of the workers, the public and the environment.

4.5 THIRD-STAGE INVESTIGATION

This is essentially a continuation of the second-stage investigation through the construction phase of a development. It is designed to determine quickly any changes that occur to the gas regime as a result of construction work or other changes in the site conditions. The location of sampling points should reflect the detail of the development.

4.6 FOURTH-STAGE INVESTIGATION — POST-DEVELOPMENT MONITORING

If gas is being produced it would be prudent to assume that structures erected on the site could be affected on a long-term basis and that monitoring may be necessary. For this reason, long-term (30+ years) post-development monitoring of sites and buildings should be considered and may be essential to:

- check that the installed gas-protection measures are working

- enable early detection of a hazardous situation arising

- confirm there is no off-site migration that may adversely affect adjacent properties.

The monitoring of gases in buildings is considered in more detail in Section 17.

4.7 SUMMARY

In circumstances where gas might reasonably be expected to be present on a site and where redevelopment of the land is being considered, an investigation to assess the gas condition will be required. Under these circumstances a preliminary investigation should be carried out. This should contain two essential elements:

1. A desk study related to the conditions of the site and its surroundings.

2. A visual assessment of the site together with some preliminary measurements of gas.

On the basis of the findings from this part of the investigation the requirements and planning of further investigative work may be assessed. This will include a more detailed search involving sub-surface sampling of the site to gather information on gas composition, concentration and emission rate. The third stage involves the continued monitoring of the site throughout the construction of the development. If gas is being produced, then long-term post development monitoring should be considered to maintain safe conditions both on- and off-site.

The scanner detects the infra-red radiation that all objects naturally emit. The amount of radiant energy given off increases with higher surface temperatures. A visual heat picture is then produced on a monitor in which the colour shading corresponds to the changes in the surface temperature of the area being scanned.

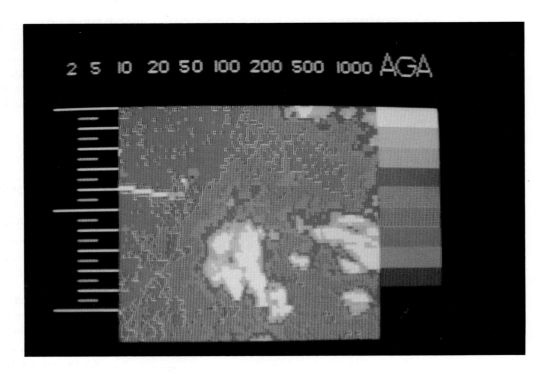

Figure 6   *An infra-red thermogram of a site emitting methane next to a borehole and gas pipeline. By permission of Wimpey Enviromental*

5 Interpreting the results of an investigation

5.1 INTRODUCTION

It may seem out of sequence to consider the interpretation and assessment of the results of an investigation before considering the means by which measurements are made and data obtained. However, interpreting the results is the most difficult part of any investigation, and an appreciation of the difficulties involved in this task is vital. It will help in defining the requirements of the investigation, ensure that appropriate measurements are obtained and assist those who use the results to assess the quality of the information and data provided.

The purposes of an investigation and, hence, the basis for the interpretation of the data are:

• to define quantitatively the conditions on the site

• to use that quantitative data to assess the need for measures to mitigate the gas problem

• to use the quantitative data as the basis for the design of protection systems should they be required.

The gas-protection system must maintain safe conditions inside the building regardless of changes in the external conditions. The essential problem in assessing the site investigation data is establishing reasonable criteria based on measurable parameters (composition, concentration and emission rate) which can be used for the design of effective gas-control systems.

Before examining this problem in more detail the following general comments are worth making:

1. For a gas which is potentially flammable and/or asphyxiating and/or toxic, between the source of the gas and clean air there will exist a mixture which is flammable and/or asphyxiating and/or toxic.

2. The flammable and/or asphyxiating and/or toxic mixture may not necessarily be at a location where it poses any threat, e.g. if a mixture in the ground is flammable it does not pose any threat provided it remains in the ground and does not extend to any other location where it may contact an ignition source.

3. When diluted to a concentration which is below the flammable, asphyxiant and toxic concentration, the gas mixture no longer represents a hazard.

These statements may appear obvious but they are the only basis on which the safety of a site and its development can be judged.

If a site is generating or contains gas, and development is proposed, the problem that needs to be resolved is not: 'Is the site safe?', but rather: 'Can additional measures be incorporated into the development to make and keep it acceptably safe for the end-user and neighbours?'

So-called trigger concentrations have been set for some types of contamination other than gases, which, if exceeded, may require some appropriate remedial action to be taken. This may be appropriate for some forms of chemical contamination where the extent of contamination can be defined and dealt with. This is not possible with gas contamination because the generation process is in a continuous state of flux. Thus, an apparently 'safe' site may, as a result of changing conditions, become 'unsafe'.

It should always be remembered that the conditions of ultimate relevance to the safety of a development are those occurring once the redevelopment programme is complete, not those at the start. This is true, and independent of, any measurements and safety precautions taken before and during the development period.

5.2 UNDERSTANDING THE MEASUREMENTS

The first problem for many people in interpreting a report of a site investigation is actually understanding the details of the measurements taken. Results of an investigation may be presented in many different formats with measurements quoted in a variety of units. For example, the concentration of methane may be given as ppm v/v (parts per million by volume), % LEL (percent lower explosive limit), or % v/v (percent volume). Ideally all measurement should be related back to a common scale, preferably % volume. There is some advantage in the use of the LEL scale to measure for flammable gases, as it is the scale that appears on many instruments. It is also directly related to the hazard and independent of the actual volume. To obtain an accurate conversion from % LEL to % volume the exact nature of the gas mixture must be known.

Figure 7 shows how these units of measurement are related for methane, although it must be realised that the % LEL values are strictly valid only for methane in air and will be different for other flammable gases and gas mixtures.

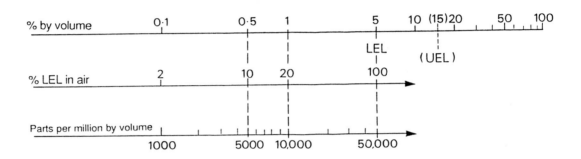

Figure 7 *Methane concentrations in air — conversion of units*

The report of an investigation should give full details of the measurements taken, i.e.:

• how — is the instrumentation used

• where — is the location, type and depth of the sampling point

• when — the date and time of the measurements

• the weather conditions prevailing at, before, and after the time of measurement.

The report should also give details of the sampling and measurement protocol followed, the calibration details of the instrumentation and the type of sampling point.

This information will help to maintain comparability should further measurements be required, help the user of an investigation report to be sure that the investigation has been carried out in a competent manner, and ensure that the results are reliable.

Figure 8 gives an example of the sort of information that should appear in the results log of an investigation report. It is not meant to represent the only way of presenting the results, but indicates what is considered to be the minimum required information.

Name of site: Date:

Sample point ref.:

Sample point location:

METEOROLOGICAL DATA

Air temperature (°C):

Atmospheric pressure:

Weather conditions (i.e. sun, rain, overcast):

Wind speed/direction:

GAS DATA

Detection instrument used:

Gas measured:

Concentration recorded:

Scale:

Time reading was taken:

Sampling method used (e.g. borehole, trial pit, probe):

Name:

Signed:

Figure 8 *Example record of gas measurement*

5.3 ASSESSING THE RESULTS

There are three important questions to ask in assessing and interpreting the results from the investigation of a gas problem:

1. How are data collected from point-source measurements related to the overall site conditions?

2. How are the data from measurements of ground conditions and emission rates related to what might migrate and appear in a building?

3. Do the data indicate that the site is acceptably safe for the development proposed?

Measurements from individual sampling points at best only give a measure of the conditions at that location at the time of measurement.

To characterise the gas regime on a site fully would require many measurements at many locations over long periods. This, in most instances, will be practically and economically impossible to achieve. However, a reasonable number of measurements made using techniques and methods described in the body of this report will, if carried out correctly, provide an adequate overall assessment of the ground-gas conditions of the site.

The relationship between the gas conditions in the ground and what might ultimately appear in a building is extremely complex. It depends on a large number of changing variables and is not well understood. With the present state of knowledge it is not possible to give definitive advice and guidance on this subject. It is only possible to say that where a potentially hazardous gas concentration is in the ground, then there is the potential for hazard to construction works on, in or near that ground unless actions are taken to prevent it.

5.4 A 'SAFE' CONCENTRATION

The question still remains: 'What is a safe concentration?'.

'Safe' conditions for inside buildings can be defined as follows:

> A safe condition is one where the concentration of a gas never exceeds its LEL or its long-term Occupational Exposure Limit (OEL) or the concentration at which the atmosphere becomes asphyxiating.

It follows that strictly the only safe site is one where the concentration of a gas, whenever and wherever measured, is found to be below these values and can be guaranteed to remain so. Even if such sites exist, the available techniques for measurement would not establish this beyond doubt, much less guarantee the continuance of safe conditions. It must, therefore, be concluded that no site which is susceptible to the presence of gases in the ground can be considered safe with absolute confidence. This being the case, an alternative approach to take is: 'What is an acceptable concentration that gives an adequate margin of safety?'.

It is not easy to define the margin of safety on any of these values (i.e. LEL, OEL, asphyxiating concentration) which makes it desirable, advisable or essential to take actions to mitigate them or their effects, and to ensure that conditions remain safe.

For flammable gases in an industrial environment an upper concentration limit of 20% LEL is often recommended. If this is set as an 'alarm' value, when the concentration of a flammable gas begins to increase, an early warning will allow action to be taken to prevent a hazard arising. In the case of gases from the ground entering a building, a similar system might be used to trigger an automatic ventilation system.

Where safety is reliant on passive protection a greater safety margin is advisable. This can be achieved with a two-level alarm system: a low alarm level, perhaps as low as 10% LEL which indicates a rising concentration but at which no immediate action is required, and a high alarm level, say 20% LEL, at which a building would be evacuated.

A similar argument may be applied to the presence of toxic gases if encountered at concentrations in excess of say 25% of their OEL. For the most commonly encountered asphyxiant/toxic gas, carbon dioxide, the long-term OEL is 0.5% by volume. Therefore a concentration in excess of 0.125% by volume of carbon dioxide would perhaps give cause for concern.

The presence of other gases or materials above 25% of their long-term OEL would also lead to the designation of a site as 'unsafe on toxic grounds'.

These 'alarm' concentrations are concentrations measured in the building itself. They are not necessarily the concentration in the ground surrounding the building. If protection measures are working effectively then concentrations of gas in the ground well in excess of these values may be tolerable if not desirable.

5.5 INTERPRETING EMISSION RATE MEASUREMENTS

The interpretation and assessment of emission rate data is even more difficult and less well understood than that of concentration measurement.

In assessing the total emission rate from the surface area of a site, an extremely pessimistic view would be to take the maximum flow rate of methane measured at any one borehole. (Note: this will depend on both the concentration and observed total volume flow rate.) This value is then corrected to give the emission rate/m² for the whole site. This may seem a drastic assumption, but two-hundredfold increases in flow rate have been observed from the same borehole from one measurement to another.

An alternative, less conservative, approach is to assume that the maximum volume flow rate from one borehole is equivalent to the flow from 10m² of the surrounding surface. This means that a single size borehole must be adopted as a standard for all measurements, if results are to be directly comparable. It also requires considerable interpolation to estimate the emission rate over the whole surface.

5.6 SUMMARY

Interpreting the results is the most difficult part of an investigation. It should seek to provide an overall picture of the gas activity on the site, so that effective gas control measures can be implemented. An investigation report should give full details of the measurements taken, the sampling and measurement protocol used, instrument calibration details and type of sampling point used.

In assessing the data three important points have to be considered:

- the relation between the data and the overall site conditions

- the relation between the data and the potential for gas entry into structures

- the adequacy of the data in indicating the degree of safety of the site for the proposed development.

There are no hard-and-fast rules for safe gas concentration levels on a site, but 20% of the LEL for flammable gases and 25% of the OEL for toxic gases are often quoted.

Various approaches can be used to select 'worse case' gas emission rates for a site, most of which make sweeping assumptions. One method takes the maximum flow rate measured at any one borehole and assumes that this is the rate for the whole site. Another way assumes the maximum flow rate from a borehole is equivalent to the flow from 10m² of the site surrounding the sampling point. Whichever approach is adopted an accurate interpretation of the data is very difficult.

6 Measuring gas – identifying the source

6.1 INTRODUCTION

Once gases have been detected on a site it is important that, if possible, their source should be determined at an early stage. There are a number of techniques available that can assist in resolving this problem: gas chromatography, mass spectrometry, radio-carbon dating, stable isotope measurement – each of these is discussed below. It is likely that positive identification may require the use of more than one of these techniques and that the unambiguous identification of a source will not always be possible.

6.2 IDENTIFICATION FROM COMPOSITION

The analysis of the composition of a gas can help to identify its source. For example, the proportion of carbon dioxide relative to that of methane typically found in bacteriogenic gases is greater than 30%, whereas in mains gas and gases of thermogenic origin it is generally less than 10%.

The measurement of major constituents of a gas mixture can be carried out on site quite reliably and with reasonable accuracy using the portable instruments discussed in Sections 9 and 10. However, the use of on-site measurement as the sole means of identifying the source of a gas is not recommended. Carbon dioxide can be easily removed from a gas by dissolving in groundwater or by reactions with other chemicals in the soil. Migration may also cause compositional change in a gas. Its composition at the sampling point can be quite different from that at source.

The most reliable method for determining gases is to take samples for laboratory analysis by gas chromatography and/or mass spectrometry. Laboratory measurements on major gases (in conjunction with on-site use of calibration gases) provide a vital check on the reliability of field measurements.

When analysing gas emissions from the ground, as well as the major constituents there are often many minor ones present (especially in landfill gas), such as hydrogen sulphide, carbon monoxide, hydrocarbons, halogenated hydrocarbons, and organosulphur compounds. Analysis of the hydrocarbon content may also prove useful, as the percentage of higher alkanes such as ethane in the gas is indicative of its origin. Generally, geologically derived methane contains up to 15% ethane and higher hydrocarbons, whereas biogenic methane only contains trace amounts.

It may also be possible to distinguish between mine gas and mains gas if the precise composition of mains gas in the local distribution network is known. This information can be obtained from the British Gas analytical service. Mains gas contains sulphides and mercaptans as odourants, and long chain hydrocarbons such as octane and nonane. Some mine gases may contain small quantities of helium.

The detection and measurement of these trace gases can therefore be useful in tracking down the source of the gas emission. But as with the major components, trace components may also be lost by chemical changes occurring in the ground during migration, by solution in groundwater or by adsorption on to clays and other minerals. It is also possible that gases such as landfill gas may have larger than expected amounts of the higher hydrocarbons if the waste in the fill contains substances that generate these gases on anaerobic breakdown by micro-organisms.

Despite the difficulties involved, a detailed analysis of the composition of a gas mixture using gas chromatography and/or mass spectrometry can be invaluable in providing evidence as to its origin. Identification of trace components is also important when assessing the potential health hazards posed by gases such as carbon monoxide and hydrogen sulphide and deciding on gas control and disposal.

6.2.1 Gas chromatography

Gas chromatography is probably the most widely used tool for gas analysis. Modern laboratory-based gas chromatographs are very sensitive, capable of determining compounds in the concentration range 0.1 parts per billion to 1 part per million. The power of the technique lies in its ability to determine qualitatively and quantitatively a wide range of compounds through the use of different separating columns and detectors. It also has the added advantage that it can be coupled to other techniques, such as mass spectrometry, and used as an extremely powerful tool to give detailed accurate analysis of gas samples. Provision for gas chromatography should be made in all gas investigations.

The operation of gas chromatographs requires specialist knowledge in the choice of separating columns, detectors and operation, if the most useful results are to be obtained. As the use of this technique can be expensive, the analysis requirements should be defined carefully before it is employed.

Gas chromatography — the basic principle behind gas chromatography is the separation of gases and vapours by passing them through a separating column on a stream of carrier gas (usually an inert gas such as helium, argon or nitrogen). The degree of separation of a mixture depends on the length of the column and the affinity to it of the various components in the gas mixture. The variety of separating columns that may be used is extensive. The choice of the type of column therefore requires special expertise and a knowledge both of the columns that are required to separate particular types of gas or vapour and of the sorts of components that may be present.

The separated gases pass through a detector to be measured. The choice of detector depends on the nature of the gases being measured. Some common examples of detector types and the gases and vapours they can be used to detect are given below.

- *Flame ionisation detectors* — used for the detection of the majority of organic compounds.

- *Thermal conductivity detectors* — used to detect most compounds, but are particularly useful for the measurement of inorganic gases and low molecular weight organics.

- *Electron capture detectors* — these detectors are specific to materials which exhibit electron capture properties, e.g. halogenated organic materials.

- *Photoionisation detectors* — used to detect most organic compounds. Do not detect oxygen, nitrogen, carbon dioxide, carbon monoxide or water vapour.

- *NICAT detectors* — a nickel catalyst system used to measure low levels of carbon monoxide and carbon dioxide by first converting them to methane. It can also be used to measure the methane content of hydrogen, nitrogen and air.

By using various combinations of separating column and detector, complex separations and analysis can be performed. For example, using photoionisation and electron capture detectors in series allows the gas chromatograph to analyse a very wide range of organic chemicals, both non-halogenated and halogenated. A gas chromatograph using this series of detectors would be very useful for determining the concentrations of trace compounds in a gas, such as volatile organic compounds (VOCs), which may present a toxic hazard.

6.2.2 Mass spectrometry

This is an extremely powerful technique which can be used to provide a very accurate and detailed analysis of samples. Mass spectrometers are generally used in conjunction with gas chromatographs, which undertake the actual physical separation of the components of the gas mixture while the mass spectrometer is used to identify the individual compounds.

It is possible to obtain portable instruments, but normally their use is limited to off-site analysis as they require a large power supply (2-3kW) and are only portable in the sense that they can be transported by van, i.e. not hand-portable. The major disadvantage with this technique is the cost as the equipment is expensive and specialist operators are required.

Mass spectrometry − this is a technique that allows the identification and measurement of a chemical by determining its mass (molecular weight) and its 'mass spectrum'. The mass spectrum is determined by the way the molecule fragments when subjected to a strong magnetic field, for each molecule has an essentially unique mass spectrum. There are several different kinds of mass spectrometers available, but the most common is the electron-impact magnetic-sector instrument.

The mass spectrum of a compound may be extremely complex and analysis is usually carried out by comparison with libraries of mass spectra using computers which form part of the instrument. The technique therefore requires specialist knowledge for both the operation and interpretation of the results.

6.3 IDENTIFICATION BY AGE AND FORMATION PROCESS

As well as identification from composition, the source of a gas may be identified if its age and the process by which it was formed can be determined. The techniques available for doing this are known as *radio-carbon* or *carbon 14 dating* (to determine age) and *stable isotope measurement* (to determine the formation process).

6.3.1 Carbon 14 dating

Carbon dating can be used to distinguish between two sources of methane − ancient geological methane (mine, natural gas) and newly-formed methane (marsh gas, landfill gas), and can also be used to determine the source of carbon dioxide.

Atmospheric carbon dioxide contains a constant proportion of radioactive isotope carbon 14 (^{14}C). While living, all organisms absorb this isotope in the same proportion, but after death absorption stops and the radioactive ^{14}C begins to decay to the non-radioactive carbon isotope, carbon 12 (^{12}C). ^{14}C decays to half its original activity after 5730 years so the proportion of ^{12}C to the residual ^{14}C indicates the period elapsed since the death of the organism. By determining the amount of ^{14}C in a gas sample, the 'age' of the methane can be estimated. A relatively large proportion of ^{14}C would indicate the source as recent, e.g. from a landfill, and little or no ^{14}C would suggest a geological origin. Carbon 14 dating techniques cannot be used to distinguish between mine and natural gas as they are so old that all the carbon 14 isotope will have decayed.

Two methods of carbon dating are currently available[18]: *radiometric dating* and *mass spectrometric dating*. The former is the more established technique, but can require a large sample which may be difficult to obtain. Mass spectrometric dating has the advantage over radiometric dating in that it requires only very small samples (minimum 1mg), and it is much quicker as the species counted are generated in the instrument and not from natural decay. A disadvantage is that not many laboratories can offer this service at present.

Radiometric dating – There are two radiometric dating techniques currently in use, both of which are well established, but require a long counting time of several weeks or even months, depending on the gas concentration. The first method involves converting the methane into benzene. A scintillant is then added and the radioactive decay of the ^{14}C in the benzene can be counted using a liquid scintillation counter. The second method involves the purified methane being used as the filling gas for a gas proportional counter. The advantage of this technique is that only milligram quantities are required whereas the scintillation counter needs gram quantities. Below these quantities it is not possible to use radiometric dating.

Mass spectrometric dating – this fairly new technique uses accelerator mass spectrometry as the basis for detection. The methane is converted to graphite which is bombarded with electrons to produce carbon ions. ^{14}C and ^{12}C ions have different masses so that when these are then accelerated through a magnetic field (which separates them), a detector can distinguish between them and count the number of ions derived from each carbon isotope.

6.3.2 Stable isotope measurements

Methane from a biogenic source, i.e. gas formed from organic matter, can be further classified as bacteriogenic or thermogenic. Bacteriogenic gas is formed by microbiological activity, either ancient or recent. Thermogenic gas is formed as a result of thermal degradation of organic matter at high temperature and pressure and is normally ancient in origin. Bacteriogenic and thermogenic methane have different proportions of non-radioactive carbon isotopes ^{12}C and ^{13}C which can be measured and used to identify the origin of the gas. The technique is still being developed and requires specialist knowledge and equipment for its operation. Interpretation of the results is very difficult and cannot be used to distinguish between recent and ancient bacteriogenic sources of gas. For a more detailed review of this technique see References 7 and 19.

The technique of stable isotope measurement compares the ratio of ^{12}C to ^{13}C in a sample to a known standard. The unit used for measurement of isotopic ratios is the delta value (δ) given in units of per mil ($^o/_{oo}$) which is defined as :

$$\delta = \frac{R(sample) - R(standard)}{R(standard)} \times 1000$$

R = isotopic ratio of the element.
Hence for $\delta^{13}C$: R = $^{13}C/^{12}C$

A positive value of δ signifies that the sample contains a greater proportion of the heavier isotope than the standard; a negative value indicates an abundance of the lighter isotope relative to the standard.

The ratio of ^{12}C to ^{13}C in the gas sample varies according to the origin of the gas. Methane of bacteriogenic origin contains a larger proportion of ^{12}C to ^{13}C than methane from a thermogenic source, and thus has typical $\delta^{13}C$ values of -60 to -100 $^o/_{oo}$. Thermogenic methane increases its proportion of ^{13}C with age and therefore has $\delta^{13}C$ values of -20 to -60 $^o/_{oo}$.

Similarly, the isotopic ratios of the oxygen isotopes ^{16}O and ^{18}O, and the hydrogen isotopes ^{1}H and ^{2}H (deuterium, D) can also be used in this way.

There are, however, a number of limitations with this technique:

1. The boundary value of -60 $^o/_{oo}$ between bacteriogenic and thermogenic methane is not always found in practice. Bacteriogenic methane has been discovered from various sources with values less than this figure.

2. Microbial oxidation of methane can cause an increase in ^{13}C, bringing it into the thermogenic range, and hence creating difficulties in distinguishing between actual thermogenic methane and oxidised bacteriogenic methane.

3. Methane from landfill contains a larger proportion of ^{13}C than bacteriogenic methane from other sources, and thus will have $\delta^{13}C$ values that are normally associated with thermogenic methane.

In view of these limitations stable isotope measurements should be undertaken in conjunction with the other techniques described earlier in this chapter.

6.4 SUMMARY

Determining the source of the gas is very important if the most effective monitoring and control techniques are to be used. There are two ways of approaching this problem: one is to determine the detailed composition of the gas (using such techniques as gas chromatography and mass spectrometry), which can be characteristic of a specific source. The other way is to determine the age and formation process of the gas using carbon dating and stable isotope measurements.

7 Measuring gas – choosing instrumentation

7.1 INTRODUCTION

Although the number of instruments available for measuring gases in the ground is large, the instruments themselves are mainly based on a limited numbers of sensors. Each type of sensor places some limitation on the use of a particular instrument. An understanding of how a particular device works therefore provides the user with an advantage in both the selection of the most appropriate instrument for a specific task and in understanding the measurements taken and how anomalous readings might arise.

Many gas sensors can be used to detect more than one type of gas, although it is sometimes possible to manufacture an instrument can be made 'single gas specific'. Table 1 lists the types of sensor used in instruments for gas measurement for the commonly encountered ground gases.

The sensors described have been used singly and in combination to provide a range of instruments which can be used for the measurement of gas in the ground.

7.2 CHOOSING AN INSTRUMENT

Before choosing any instrument for ground gas measurement there are some basic questions which should be considered. These concern the instrument(s) which might be used and the job required of them. On the basis of the answers to these questions the most appropriate instrument and measurement method can be chosen.

- *what do I want to measure?* This may seem obvious, but without a clear idea of what gases are to be detected, what accuracy and what resolution are required, it is easy to make the wrong decision about what instruments to use

- *can I take the measurements myself?* Measurements of gas in the ground should be made only by competent and trained personnel. Most instruments are comparatively easy to use and measurements can be carried out by a well-trained technician without difficulty. However, some measurements (e.g. gas chromatography and mass spectrometry) are much more specialised techniques and will require additional expertise. If in doubt, all measurements should be made by people familiar with the procedures required

- *can all the measurements required be made with a single instrument?* It will be an advantage both in the amount of equipment that has to be carried and in the ease of measurement if a single instrument capable of taking the necessary measurements can be found. This will not always be possible

- *is the instrument robust?* For on-site measurements an instrument will need to be sturdy and well constructed to resist damage during handling and from adverse weather conditions

- *does the instrument have an independent power supply?* For on-site work an independent power supply is essential. Rechargeable batteries should have a life between recharging of at least six hours continuous use under site conditions

- *how heavy and what size is the instrument?* An instrument that is heavy or cumbersome may be difficult to carry around a site

- *does it have its own sampling pump?* It can be an advantage if an instrument has its own internal sampling pump, although these can lead to very short battery life. Alternatives are hand-aspirated sampling systems or independent pumps with their own power supply

- *what type of display system is used?* A clear readable display, whether analogue or digital, is important. This is particularly so for instruments which use the same display for multiple ranges or different gases

Table 1 Sensors of instruments for gas measurement

Instrument	Gas detected	Typical range of sensitivity
Infra-red	Methane	0 - 100% volume
		0 - 10% volume
	Carbon dioxide	0 - 100% volume
		0 - 10% volume
		0 - 5% volume
	Carbon monoxide	0 - 100% volume
Catalytic	Flammable	0 - 100% LEL
	Hydrogen sulphide	0 - 100 ppm
Thermal conductivity	Flammable	0 - 100% volume
	Carbon dioxide	0 - 100% volume
Catalytic/thermal conductivity	Flammable	0 - 100% LEL
		0 - 10% LEL
		0 - 100% volume
Flame ionisation	Flammable	0 - 30000 ppm
		0 - 10000 ppm
Electrochemical	Oxygen	0 - 100% volume
		0 - 25% volume
	Hydrogen sulphide	0 - 1000 ppm
		0 - 500 ppm
		0 - 25 ppm
	Carbon monoxide	0 - 4000 ppm
		0 - 2000 ppm
		0 - 100 ppm
		0 - 25 ppm
	Hydrogen	0 - 2000 ppm
		0 - 1000 ppm
		0 - 200 ppm
Paramagnetic	Oxygen	0 - 100% volume
Semiconductor	Flammable	10 - 40% volume
		0 - 100% LEL
		0 - 20% LEL
		25 ppm - 1% volume
	Hydrogen sulphide	0 - 200 ppm
		0 - 100 ppm
		0 - 50 ppm
	Carbon monoxide	50 - 500 ppm
		20 - 100 ppm
Detector tubes	Hydrocarbons (methane)	0.05 - 2.4% volume
		100 - 3000 ppm
	Carbon dioxide	2.5 - 40% volume
		0.5 - 20% volume
		0.13 - 6% volume
		100 - 11500 ppm
	Oxygen	6 - 24% volume
	Hydrogen sulphide	1 - 40% volume
		0.1 - 4% volume
		10 - 3200 ppm
		1 - 240 ppm
		0.1 - 4 ppm
	Carbon monoxide	1 - 40% volume
		0.05 - 4% volume
		2.5 - 2000 ppm
	Hydrogen	0.5 - 2% volume
		0.05 - 0.8% volume
Photoionisation	Hydrogen sulphide	0 - 2000 ppm
		0 - 200 ppm
		0 - 20 ppm

Note: Typical ranges of sensitivity have been obtained from manufacturers' literature at the time of writing and are not necessarily the only ranges available.

- *are the operating instructions clear and precise?* If the instrument is to be used effectively this is essential. The instructions should also contain a detailed instrument specification, calibration and checking procedures, recommended service and calibration intervals, and details of the accuracy and stability of the instrument under different conditions. As at present there is no independent certification standard for instruments to be used specifically for measurements of gases in the ground, the user should establish with the manufacturer that the conditions under which the instrument is to be used will not adversely affect the readings

- *are the controls clear and easy to use?* The controls ought to be clear and simple, particularly for instruments which measure multiple gases and have multiple ranges. There should be no confusion over the gas being measured or the range of detection

- *does the instrument have an internal memory and computer logging facility?* Although not essential, the ability to store internally the data that are recorded and to be able to transfer that information to a computer for long-term storage is extremely useful, and a facility that is being increasingly offered on many new instruments. It would be an essential feature of any system that was to be used for long-term monitoring

- *is the instrument intrinsically safe?* If the instrument is to be used in confined spaces where flammable gases may be present, the instrument could be a potential ignition source unless it has been certified as being intrinsically safe

- *with what mixture was instrument calibrated?* It is important to know what the factory calibration mixture was. For example, thermal conductivity devices can be calibrated to detect methane in air, methane in a carbon dioxide/air mixture, methane in carbon dioxide, or carbon dioxide in methane, etc. It is important that the instrument calibration is checked with an appropriate mixture. Instruments that are used in mixtures other than those for which they have been calibrated will not give accurate readings, and these could under some circumstances be significantly in error

- *how frequently should the instrument be serviced and what is the cost?* Some manufacturers recommend an annual service, others a biannual service. The prices charged will obviously differ between manufacturers

- *does the instrument's sensor have a limited normal life?* Some sensors will only operate accurately for a limited lifetime (e.g. one to two years) before replacement is essential. Advice from the manufacturer should be sought

- *could the life of the sensor be reduced by components of the sampled environment?* Some sensors may be damaged by poisons or exposure to high concentrations of gas. Advice from the manufacturer should be sought

- *what customer support does the manufacturer offer?*

8 Measuring gas — a sampling methodology

8.1 INTRODUCTION

It is good practice to lay down a sampling methodology or protocol which is used for all measurements taken. In this way possible differences in results caused by the use of inconsistent sampling methods and different operators can be minimised to ensure the comparability of results taken at different times. The following is presented as a reasonable measurement protocol; it is not meant to be definitive, as particular circumstances may dictate that alternative actions are taken. Whatever the measurement methodology or protocol chosen for a particular measurement, it should be followed consistently. If it is necessary to change the protocol, e.g. because of site conditions, it should be clearly noted in the report of the investigation.

The following protocol is given as a check list of actions that should be taken in all circumstances. The actions are dependent upon answers to questions given in the text, which in turn will depend on particular circumstances. This is followed by a decision tree that attempts to summarise this information. The protocol given applies to the measurement of gas composition and concentration with portable instruments.

8.2 PRE-SITE CHECKS

1. Check general condition of instrument — pumps, sample tubes, power supply, displays.

2. Check service and maintenance records to ensure that any maintenance and servicing required has been carried out according to the manufacturer's instructions.

3. Check instrument calibration.

4. Record details of instrument, faults found, remedial action taken, satisfactory calibration.

8.3 DAILY ON-SITE CHECKS

1. Check general condition of instrument — pumps, sample tubes, power supply, displays.

2. Record details of instrument, any faults found, remedial action taken.

3. Check calibration at least at the beginning and end of every day. If large numbers of readings are being taken it is advisable to check the calibration at regular intervals throughout a day. Record date and time of each calibration, noting particularly when adjustments are required if the calibration has drifted.

8.4 ACTIONS AT EACH SAMPLING POINT

1. Record date, time, sample point type and location.

2. Check depth of water table, and estimate volume of gas available for sampling within the sampling point.

> **Question.**   Is the available sample volume adequate to take all measurements required, i.e. separate samples for oxygen, methane, carbon dioxide, other gases?

To answer this question the following should be considered:

Does the instrumentation to be used destroy the sample in measurement?

The sample point volume should be at least 10 times the volume of the instrument used including pump and external tubing.

Is the sampling point replenished from the surrounding ground readily so that sampling does not significantly affect the sample itself?

This might only be discovered by trial and error.

Yes. Proceed with measurements

No. Consider:

   (a) removing single sample for off-site analysis (Section 12.4).

   (b) use of *in-situ* sensor (Section 16.4) that does not destroy sample.

   (c) recirculation technique (Section 12.3) that does not destroy sample.

Question. **Do any of the instruments being used require oxygen for accurate measurement?**

Measurements using catalytic sensors and flame ionisation detectors require a minimum amount of oxygen.

Yes. Measure oxygen concentration first

No. Proceed with measurements

3. Take measurements on least sensitive scale first.

4. The external sampling volume should be flushed until a constant reading is obtained or at least for three to five full volume changes.

5. Record measurement.

6. Repeat measurement, on more sensitive scale if necessary.

7. Record measurement.

8. Repeat measurement again if there is a significant difference between first and second measurements.

9. Record maximum, mean and variation in measurements obtained.

10. Record date, time, temperature, weather conditions.

11. Record depth at which samples were taken.

12. Repeat procedure for measurement of other gases.

13. Consider need to take sample for off-site analysis as a quality control measure for on-site measurements.

It is advisable to take a representative sample of gases from some sample points for off-site analysis. This can be used to obtain a more comprehensive determination of the composition including trace gases, and provide a quality control check on the measurements made on site.

8.5 SUMMARY

A sampling methodology is necessary if comparable results are to be obtained when taking measurements at different times. Daily checks on the instrumentation should be made both before going on-site and while at the site.

Compare, before leaving the site, current readings with previous data. If possible use more than one system of measurement.

A list of actions to be taken and questions to be answered when taking measurements should be drawn up. This should include, as a minimum, recording the location of the sampling point, date, time, weather conditions, depth of the water table within the sampling point, and ensuring the sample volume is adequate for the necessary readings. The necessity for oxygen concentration measurements and the need for off-site analysis of the gas should also be considered.

9 Measuring methane

9.1 INTRODUCTION

Methane gas can be measured and detected using a variety of instruments with different types of sensors. The most commonly used instruments for on-site measurement are fitted with detectors operating on one or more of the following types of sensor principles:

- infra-red absorption

- catalytic oxidation

- thermal conductivity

- flame ionisation.

Other detection techniques, but which are not widely used in the context of a site investigation, are:

- semiconductor detectors

- chemical indicator tubes.

Table 2 summarises the typical concentration range for which instruments utilising each of these sensors can be used, together with typical values for the lower limit of sensitivity which may be expected from the instrument when operated under *ideal* conditions.

Table 2 Methane instruments

Instrument	Range	Sensitivity	Application
Infra-red	0 - 100% vol 0 - 10% vol	±3% vol ±1% vol	Suitable for monitoring methane and carbon dioxide (see Table 3).
Catalytic	0 - 100% LEL 0 - 10% LEL	±2% LEL ±0.5% LEL	Suitable for monitoring low levels (<5% vol) of flammable gas if sufficient oxygen is present.
Thermal Conductivity (usually with catalytic)	0 - 100% vol	±2% vol	Suitable for monitoring methane and carbon dioxide (see Table 3), but must be calibrated with required gas.
Flame ionisation	0 - 1% vol	±0.5ppm	Suitable for detecting low levels of flammable gases. MUST NOT be used in confined spaces. It should be used in conjunction with other detectors.
Semiconductors	0 - 20% LEL		Capable of detecting combustible gases at very low concentrations.
Detector tubes	0.01 - 2.4% vol		Suitable for detecting hydrocarbons and other gases (see Tables 3 and 4) at very low concentrations.

Note: Typical ranges of sensitivity have been obtained from manufacturers' literature at the time of writing and are not necessarily the only ranges available.

9.2 INFRA-RED ABSORPTION

Methane (along with many other gases) can be detected and measured with instrumentation fitted with infra-red detectors.

For any molecule which absorbs radiation in the infra-red region of the electromagnetic spectrum it is possible to construct an instrument which can measure the concentration of that gas in the sampled atmosphere. This is achieved by careful selection of a unique infra-red wavelength for measuring the gas. In practice, however, because the infra-red spectra of different gases (particularly organic hydrocarbon gases and vapours) often overlap, the selection of a unique measuring wavelength is difficult, and the specificity of an instrument to a particular gas may be reduced by interference. This problem can largely be overcome by designing selective detector systems based either on the Luft cell [20] or on selective filters so that highly specific analysers are available for many gases.

Because of their selectivity and sensitivity, a range of infra-red analysers exist which allow the measurement of individual components: as individual instruments for single gases such as methane and carbon dioxide alone; as one instrument for the measurement of binary gas mixtures such as methane and carbon dioxide; or as a single instrument for the analysis of complex mixtures of organic vapours.

An instrument based on infra-red measurement has the following advantages.

1. The detection method does not consume any component of the sample, thus allowing options such as, recirculation of the sample (Section 12.3) to be used even on small volumes of material.

   This is very useful when the amount of sample is limited. By recirculating it through the analyser more accurate measurements may be obtained.

2. The sampling cell can be made small enough to fit in the sampling point so that measurements can be made without physically removing the sample.

3. The detection system does not require the presence of oxygen.

   It is still advisable, though not essential, to determine the oxygen concentration to obtain a full picture of the site conditions.

4. The detection system can be made essentially specific to a particular gas, significantly reducing the effects of cross-interference from other gases on the sample.

   Cross-sensitivity may still be a problem in the measurement of methane if other hydrocarbon gases are present.

5. The output from many sensors can be linked to form a network to provide continuous long-term monitoring.

6. The sensor cannot be poisoned.

7. Sensors operating with different infra-red wavelengths can be incorporated into a single sensing head to allow simultaneous measurement of more than one gas.

   An example of where this has proved useful is in applying dual detectors for the measurement of methane and carbon dioxide simultaneously.

An instrument based on infra-red measurement has the following disadvantages.

1. The selection of a unique measuring wavelength, particularly for hydrocarbons, is difficult.

   The infra-red spectra of some gases and vapours, especially hydrocarbons, often overlap and this can create difficulties in selecting a measuring wavelength that is unique to the gas of interest.

2. The presence of water in the gas, particularly if it condenses in the sample cell, will affect the readings.

   Water has a very broad band of absorption in the infra-red region of the spectrum. The infra-red radiation will also be absorbed by any water present in the gas, particularly if allowed to condense within the sampling cell. This is usually overcome by heating the sampling cell to prevent condensation forming.

3. A dirty sampling cell can cause the instrument to under-read.

   Dirt on the windows of the detector cell can cause the signal given for a particular concentration to decrease, thereby registering a lower concentration than the true value.

9.2.1 Laser detection systems

LIDAR (Laser Radar) and DIAL (Differential Absorption LIDAR) systems are essentially an extension of the infra-red type of instrument in which a tunable laser is used to provide an intense narrow-band source of infra-red radiation at a selectable wavelength.

A diode laser monitoring system built around a solid state diode laser has been developed[21] that can, in principle, emit radiation in any portion of the mid-infra-red region, i.e. from 3 to 30 microns in wavelength. The system is mounted in a modified van, from which the laser beam emerges via a steerable periscope. The laser beam is directed through the portion of the atmosphere to be measured until it hits a suitably sited reflector which returns the beam back along its original path for measurement by a detector close to the laser. The amount of target gas in the chosen atmosphere is measured in principle by observing the reduction in laser beam intensity due to absorption by the gas.

As well as being able to locate the main methane emission areas of a site, this system can also determine the total flux of methane emitted.

A low cost portable system for both landfill and industrial applications is currently being developed.

The advantages of using these systems are:

1. It is possible to monitor continuously large areas.

2. They can provide immediate information on changes in methane emission patterns before, during and after development work is carried out.

3. They can be used to scan across the surface of the site to monitor gas emissions without disturbing the ground.

The disadvantages of these systems are:

1. They are very expensive.

2. They only measure the surface gas concentrations. Other methods will still be necessary to obtain valuable information of gas concentrations in the ground.

9.2.2 Ultra-sensitive detectors

A further development in the use of infra-red for the detection and measurement of methane is the analysis, by means of ultra-sensitive detectors, of the infra-red absorbed by methane from natural light. Until recently this has only been possible using detectors operating in the visible or ultra-violet regions of the spectrum, and therefore only of use in detecting gases such as ozone, sulphur dioxide and nitrogen dioxide. Now detectors have been developed which, it is claimed, make it possible for the system to work in the infra-red part of the spectrum where methane has its absorption spectra. The detector, called Otim, is based on the same principles used to process radio frequency signals, which accounts for its success at detecting gases down to one hundred parts per billion (10^9). It is hoped that a hand-held system rather like a video camera can be developed, with the advantage that it will have a wider field of view than the active laser monitor previously mentioned.

9.3 CATALYTIC INSTRUMENTS

These are probably the most widely used type of methane gas detector because of their relatively low cost and ease of operation. Although usually calibrated for methane in air for the 0-100% LEL range (0-5% by volume) this type of instrument will detect any flammable gases (e.g. hydrogen and organic vapours).

Catalytic sensors detect the presence of a flammable gas from the heat generated by the oxidation of the combustible material in a sample on a small heated sensing element. The sensing element, usually a platinum resistance thermometer, contains within it the catalyst to promote oxidation and forms part of a Wheatstone bridge circuit. An out-of-balance voltage is produced by the rise in temperature upon oxidation which is sensed and displayed either visually or audibly by an alarm when a pre-set level is exceeded.

The signal produced is roughly proportional to the concentration of the flammable gas expressed as a fraction of its LEL. *This type of sensor will detect any flammable gas or vapour and cannot distinguish one from another.*

Although the response of the instrument will be similar for any flammable mixture in terms of the % LEL measured, the absolute concentration will differ for different gases. Therefore, to accurately convert the LEL measurement of a single flammable gas in air mixture to a concentration in % by volume, the instrument should preferably only be used and calibrated with the gas it is intended to measure. Alternatively, an instrument may be calibrated with a single flammable gas, e.g. pentane or acetone, and the manufacturer's conversion tables used to correct for different gases.

General manufacturing standards and operating guidelines have been produced for this type of instrument and an investigator is advised to study these[22].

The advantages of this type of sensor are:

1. It will detect the presence of any flammable gas.

   Because the output of the sensor is based on the degree of flammability of the mixture, the sensor will indicate that a hazardous situation is arising almost regardless of the mixture composition, providing sufficient oxygen is present.

2. It is capable of detecting low concentrations of gas (typically 2% LEL).

3. It can be made easily into a portable instrument.

The disadvantages in using catalytic sensors are:

1. The sensor is not gas specific.

2. The sensor requires oxygen.

   If reliable readings are to be obtained, sufficient oxygen must be present to oxidise all the flammable gas present. It is recommended that the oxygen concentration in the sampled atmosphere should not fall below 12% by volume. If sampling from a potentially oxygen-deficient atmosphere (e.g. from a sub-surface sampling point) the oxygen concentration must be determined before measurement of methane is attempted.

3. The sensor may become saturated.

   If the sensor is used for prolonged measurement at high concentration a greater fraction of the area of the catalyst is used in the reaction, resulting in the signal levelling out as shown in Figure 9.

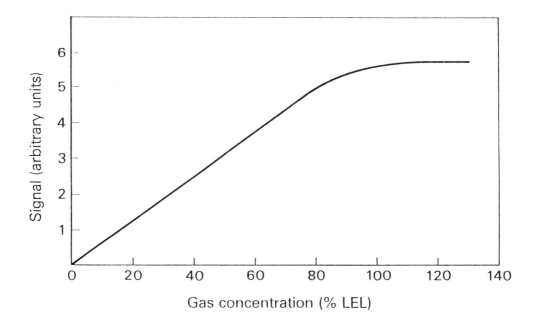

Figure 9 *Signal concentration curve for catalytic sensor*

4. In producing an output the sensing element destroys the sample (by oxidation).

5. At very high concentrations this type of sensor will read zero.

   In its crudest form this type of sensor will read less than 100% LEL for concentrations above the Upper Explosive Limit (UEL), due to insufficient oxygen being available to oxidise the flammable gas. This is a potentially dangerous feature which can be overcome by coupling the catalytic sensor with the output from a thermal conductivity sensor (see Section 9.4).

6. Catalyst poisoning may de-activate the sensor.

7. Sensor deterioration may occur with age.

   Deterioration of the sensing element could result from age or catalyst poisoning. Hence the instruments should always be checked with a reference gas before and after use.

> *Catalyst poisoning*. This occurs when mixtures of gases which contain compounds of silicon, lead, phosphorus or halogens contaminate the catalyst surface, reducing or destroying completely its activity. This potentially serious problem can be overcome by the use of poison-resistant catalysts or by the use of suitable filters which remove the damaging compounds before they can contact the catalyst. Advice from individual manufacturers should be sought on the susceptibility of a particular instrument to this form of attack.

9.4 COMBINED CATALYTIC/THERMAL CONDUCTIVITY DEVICES

The majority of commercially available catalytic type gas detector instruments are now combined with a thermal conductivity sensor. This allows the addition of an extra measuring range (0-100% by volume).

> *Thermal Conductivity Sensors* - thermal conductivity is a measure of the efficiency with which materials conduct heat. Sensors operate by measuring the difference in thermal conductivity of the sample stream relative to air (or other reference gas). This is achieved by measuring the out-of- balance voltage from a Wheatstone bridge circuit containing a sensing and a reference element. The voltage is generated by the change in resistance caused by the change in the rate of heat loss as the sample passes over a heated sensing element in the bridge. A signal can be produced which is positive for all concentrations of a gas in the sample from 0-100% by volume.

Some gases have a higher thermal conductivity than air, some a lower thermal conductivity. The measurement is therefore very dependent on the composition of the gas mixture and the reference gas (normally air).

The advantages of adding the thermal conductivity sensor are:

1. It can be used to determine the concentration up to 100% by volume of any gas present in the reference gas (normally air).

2. It can be easily combined with other detectors.

The main disadvantages with this type of thermal conductivity sensor are:

1. It cannot distinguish between gases or gas mixtures which have the same thermal conductivity.

2. The presence of other gases with different thermal conductivities can influence the readings in some instruments.

3. Errors can occur at low gas concentrations.

The manner in which the thermal conductivity detector is combined with the catalytic sensor is critical. The signal from the catalytic sensor and the thermal conductivity sensor may be separate (apart from sharing the same output meter), in which case the instrument will have the limitations and advantages of both types of sensor. Alternatively the output from the thermal conductivity sensor may be electrically coupled with that of the catalytic sensor so that on the LEL range the reading will always be off-scale if the concentration is above the LEL. This offers the advantage that on the LEL range the possibility of ambiguous readings being obtained is reduced and the thermal conductivity sensor may be used independently to directly measure concentrations above the LEL up to 100% v/v.

In addition to the comments made above about each type of sensor, the following are points to note when using these combined or coupled devices:

1. It is important to establish that sufficient oxygen is available for the correct operation of the catalytic sensor. It is therefore strongly advised that an oxygen concentration measurement is made before attempting to measure methane.

   Some instruments will also incorporate an electrochemical oxygen sensor (see Section 10.3.1) for the determination of the oxygen concentration.

2. The initial measurement of methane should always be made on the least sensitive range available, i.e. 0-100% volume, followed by measurements on more sensitive ranges, 0-100% LEL, 0-10% LEL.

3. Because the sensing method destroys the sample, problems may be encountered in obtaining a reliable reading where the sample is not sufficiently large.

   If the sample is insufficient to allow a full sequence of measurements to be made (i.e. oxygen measurement followed by a high concentration check and then a low concentration check), interpret the results with caution. It is advisable to use either an in-situ *method of measurement (see Section 16.4) or to take the sample for off-site analysis by gas chromatography.*

4. The user should be aware of whether or not the two sensing systems are electrically coupled as this may influence the reading on the LEL range.

5. On the 0-100% volume range any significant amount of carbon dioxide may produce an apparently low reading for methane because of differences in their thermal conductivities.

6. Differences in thermal conductivities may also be reflected in a combined catalytic/thermal conductivity signal on the 0-100% LEL range.

7. With the % volume scale, the user should ensure that the instrument is calibrated for the required gas and be aware that the presence of other gases (e.g. carbon dioxide) may influence the readings.

8. These instruments generally use the thermal conductivity sensor on the % volume scale and the catalytic sensor on the % LEL scale, hence the user should be aware that on switching from the volume scale to the LEL scale a sufficient amount of oxygen must be present to obtain an accurate reading.

9. By coupling the two detectors the influence of various interference factors encountered by single devices is reduced.

10. In portable form these instruments are relatively cheap and simple to operate.

The type of instruments described above are probably the most widely used for the general detection and measurement of methane and other flammable gases. They offer a wide range of sensitivity. Although not suited for sniffing the open air (where the concentration of gas may be very low), on gassing sites, they have found considerable use in enclosed atmospheres, in sampling from surface cracks and fissures, and from purposely installed sampling points.

This type of detection system can also be used as the basis for a multi-point fixed system as well as an easily portable device.

9.5 FLAME IONISATION INSTRUMENTS

The flame ionisation instrument is a portable version of the type of detector commonly used in gas chromatographs for the detection and measurement of organic compounds. It is more sensitive than the catalytic devices, capable of measuring as little as 0.5 ppm, but has an upper limit to its detection which relies on sufficient oxygen being present to allow efficient combustion.

Flame ionisation instruments. A portable development of the type of detector often used in gas chromatographs for measuring organic compounds. The principle of this sensor is based on recording the increase in ions produced when a combustible material passes through a hydrogen/air flame, forming an electrical current. The amount of current produced depends upon the concentration of flammable gas present.

Most instruments of this type use the air drawn in with the sample to provide the oxygen to support the flame in the sensor. They therefore require a minimum amount of oxygen in the sample to operate correctly, and will not be suitable for sampling regions where oxygen may be in low concentration or absent.

This requirement for oxygen also puts an upper limit on the detection of methane, although this has been overcome to some extent by the provision in some instruments of a diluter which adds air to the sampled gas stream (normally in the ratio 10:1).

As the sensor incorporates a flame these instruments should not be used in an explosive atmosphere (or where such an atmosphere might be present but has not been defined), unless they have been certified as intrinsically safe. Their sensitivity, however, makes them suitable as a primary investigating device for sniffing the open air above gassing sites. They are generally used in conjunction with other gas detectors such as catalytic instruments.

The absence of gas as determined by this method does not necessarily imply safe conditions for all future circumstances or users, while the presence of gas certainly indicates the need for further investigation. The principal use of this type of instrument is in the early stages of an investigation or where high sensitivity is required. In the open air, it can be used to discover where gas is emitting from the ground; in surveys inside buildings it can provide an early indication of a problem. Hence its main function is to give warning and to locate areas where future measurements should be taken.

The advantages of this type of instrument are:

1. It is very sensitive.

2. It can be turned relatively easily into a low-resolution portable gas chromatograph.

   The effective use of gas chromatography with these devices requires experience and understanding and should not be attempted by unqualified personnel.

The user should also be aware of the following disadvantages:

1. It will respond to the presence of any combustible material.

2. It must have sufficient oxygen to allow efficient combustion of the sample, therefore the oxygen concentration should be determined before attempting to measure methane with this type of instrument.

3. It has a limited detection range, although this can be extended by controlled dilution of the sample with air.

4. The sensing method destroys the sample and therefore caution should exercised if attempting measurements from sampling points where the amount of sample available is limited.

5. The sensor is a potential ignition source and should only be used in open areas where the atmosphere has been determined and is not explosible.

9.6 SEMICONDUCTOR GAS DETECTORS

In recent years attention has been paid to the development of flammable gas sensors based on semiconductors.

Semiconductor gas detectors. In semiconductor gas detectors the electrical resistance of a small solid-state semiconductor is changed by the adsorption of a gas on to its surface. This change in resistance is related to the concentration of combustible gas in the atmosphere being sampled and is displayed on a suitable meter.

The advantages of sensors of this type are:

1. They are not easily poisoned by most of the gases which affect catalytic sensors.

2. They have good long-term stability and long sensor life, because of their low operating temperatures.

3. They can be made very sensitive.

Their main disadvantage has been that they are difficult to make selective, and there are few examples of practical instruments available based on this type of sensor.

9.7 CHEMICAL DETECTOR TUBES

Methane is just one of a wide variety of gases that can be determined using chemical detector tubes (Figure 10).

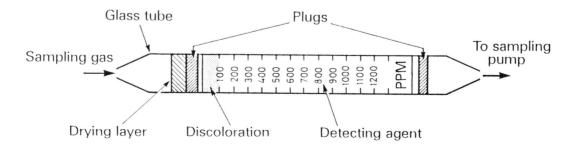

Figure 10 *Detector tube*

Chemical detector tube. A gas sample is drawn through a tube containing one or more reactive chemicals and a chemical reaction occurs, producing a colour change in the reagents. The concentration is determined either by the length of coloured stain or the intensity of colour.

There are two types of gas detector tube: pumped tubes, where the gas is drawn in by a sampling pump (Figure 11); and diffusive tubes, where the gas enters by diffusion.

The system using pumped tubes involves taking the correct tube for the gas to be measured, breaking open the ends using the tip breaker on the pump and then connecting the tube to the pump. The gas sample is then drawn into the tube and a colour change will occur, the length of which corresponds to the gas concentration.

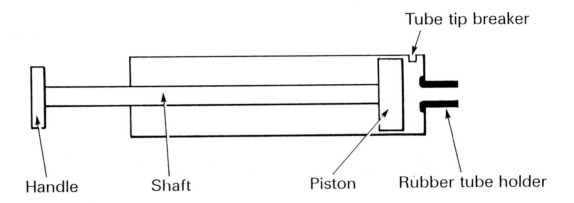

Figure 11 *Schematic diagram of a gas sampling pump*

The precision and readability of the tube depend on the reacting agents. The range of detection is normally in the region of 0.01-30% by volume. British Standard BS5343[23] sets out minimum requirements for these devices and the user of these tubes is advised to read it. Users should always follow carefully the manufacturer's instructions and consult the manufacturer about any known or suspected interfering gases.

10 Measuring other gases

10.1 INTRODUCTION

Although methane is the most common hazardous gas emanating from the ground, it is rarely found alone. Many other gases may be present such as carbon dioxide, carbon monoxide, hydrogen sulphide and hydrogen. The measurement of each of these gases is important because of their asphyxiating/toxic/explosive properties. Oxygen will also be present from the atmosphere. It is important to measure this too as some instruments rely on a minimum oxygen concentration in order to function accurately.

10.2 CARBON DIOXIDE

The detection and measurement of carbon dioxide is usually accomplished by means of an instrument incorporating one of three types of detector. These are:

- infra-red

- thermal conductivity

- chemical dectector tubes.

Table 3 lists the most common types of instrument and their application, together with their typical range of operation and lower limit of sensitivity.

Table 3 Carbon dioxide instruments

Instrument	Range (volume)	Sensitivity (volume)	Application
Infra-red (see also Table 2)	0 - 100% 0 - 10% 0 - 5%	±3% ±1% ±0.1%	Suitable for monitoring carbon dioxide in the presence of methane and air.
Thermal conductivity (see also Table 2)	0 - 100%	±10%	Suitable for use in the presence of methane and air. MUST be calibrated for gases required.
Detector tubes (see also Tables 2 and 4)	0.01 - 40%	±0.01%	For measuring low levels of carbon dioxide in gas mixtures. May be affected by hydrogen sulphide.

Note: Typical ranges of sensitivity have been obtained from manufacturers' literature at the time of writing and are not necessarily the only ranges available.

10.2.1 Infra-red instruments

Infra-red instruments offer what is probably the most reliable method of on-site measurement for carbon dioxide currently available. The advantages and limitations of the instrument are essentially as indicated for methane determination by this technique, although carbon dioxide instruments are less likely to be subject to interference from other gases as selection of a unique infra-red measuring wavelength for carbon dioxide is easier.

A description of infra-red detection is given in Section 9.2 relating to the measurement of methane.

An instrument based on infra-red measurement has the following advantages:

1. The detection method does not consume any component of the sample, thus allowing options such as recirculation of the sample (Section 12.3) to be used even on small volumes of material.

   This is very useful when the amount of sample is limited. By recirculating it through the analyser more accurate measurements may be obtained.

2. The sampling cell can be made small enough to fit *insitu* in the sampling point so that measurements can be made without physically removing the sample.

3. The detection system can be made essentially specific to carbon dioxide.

4. The output from many sensors can be linked to form a network to provide continuous long-term monitoring.

5. The sensor cannot be poisoned.

6. A single instrument with detectors operating at different infra-red wavelengths can be used to measure both carbon dioxide and methane simultaneously.

The disadvantages of the infra-red sensor for the measurement of carbon dioxide are:

1. The presence of water in the gas, particularly if it condenses in the sample cell, will affect the readings.

   Water has a very broad band of absorption in the infra-red region of the spectrum. The infra-red radiation will also be absorbed by any water present in the gas particularly if allowed to condense within the sampling cell. This is usually overcome by heating the sampling cell to prevent condensation forming.

2. A dirty sampling cell can cause the instrument to under-read.

   Dirt on the windows of the detector cell can cause the signal given for a particular concentration to decrease, thereby registering a lower concentration than the true value.

10.2.2 Thermal conductivity instruments

Because many gases have a unique thermal conductivity, instruments based on a thermal conductivity sensor can be manufactured to determine the concentration of a wide variety of single gases (e.g. carbon dioxide) in air (or other reference gas). But since the sensor only measures the difference in thermal conductivity between the reference gas and the sampled gas, the presence of multiple components in the mixture can seriously affect the readings. This can be overcome by using one of two techniques:

1. To remove component gases.

   For simple mixtures containing no more than three major components (e.g. methane and carbon dioxide in air) it is possible to take a combined measurement on the whole sample, and then a second measurement with one of the gases removed (e.g. carbon dioxide removed on a suitable filter), and by calculation determine the true concentration of both the methane and the carbon dioxide in the original sample. This method is not recommended for site work as it leaves too much room for error.

2. Use of gas chromatography.

For very complex mixtures gas chromatography can be used to separate the individual components in a mixture, and each one can then be determined separately with a single detector calibrated for each component.

The principal drawbacks of instruments which are based solely on this type of sensor arise from their non-specificity which can lead to erroneous readings under quite common conditions.

For a more detailed description of the thermal conductivity sensor see Section 9.4 on methane detection using combined catalytic/thermal conductivity detectors.

10.2.3 Chemical detector tubes

The chemical detector tube (described in Section 9.7) can also be used to determine carbon dioxide. A gas sample is drawn through a tube containing one or more reactive chemicals and a chemical reaction between the gas or vapour being sampled occurs producing a colour change in the reagents. The concentration is determined either by the length of coloured stain or the depth of colour.

10.3 OXYGEN

There are three main types of detector used to monitor oxygen (see Table 4). These are:

- electrochemical cells

- paramagnetic cells

- detector tubes.

Table 4 Oxygen instruments

Instrument	Range (volume)	Sensitivity	Application
Electrochemical cell	0 - 100% 0 - 25%	±0.5%	Monitoring oxygen content in gas mixtures, e.g. landfill gas.
Paramagnetic cell	0 - 100%	±2.5%	Monitoring oxygen content in gas mixtures. Mainly used in process control.
Detector tubes	2 - 24%	±2.0%	In areas where flammable gas present. May be effected by other gases, e.g. Hydrogen sulphide.

Note: Typical ranges of sensitivity have been obtained from manufacturers' literature at the time of writing and are not necessarily the only ranges available.

10.3.1 Electrochemical cells

Most commonly used for the determination of oxygen, instruments based on electrochemical cell sensors are available for a wide range of gases and can be used to determine the presence of carbon dioxide, hydrogen sulphide, and carbon monoxide.

They can be supplied as individual instruments or are sometimes combined within other gas detectors to provide a means of measuring the oxygen concentration and ensuring it is adequate for the detector to function.

The disadvantages with this type of sensor are:

1. The electrolyte is used up during the reaction and hence the cell has a limited life span.

> *Electrochemical cell.* The basic electrochemical device consists of two electrodes in an insulated container, fitted with a membrane that is permeable to the gas being sampled but that will retain the electrolyte (usually in the form of a gel) in the cell. With a potential applied across the cell and gas diffusing through the membrane into the electrolyte a chemical reaction takes place, producing a current which increases with the concentration of gas present.
>
> By careful design and choice of electrolyte these sensors can be made very specific to a single gas. Because the process is controlled by diffusion of gas through the membrane some temperature dependence of the measurement may be observed, although this can be compensated for electronically or by other modifications in the design.

2. They can be poisoned by certain gases.

10.3.2 Paramagnetic instruments

Apart from the instruments based on electrochemical cells, the most commonly used instruments for measurement of oxygen are 'paramagnetic' analysers. These effectively detect the partial pressure of oxygen present in the sampled atmosphere by withdrawing a sample into the instrument from which a direct reading of the concentration is obtained.

> *Paramagnetic instruments.* The detection of oxygen with paramagnetic sensors relies on oxygen's property of positive magnetic susceptibility, i.e. it causes a greater concentration of magnetic force within itself than in the surrounding magnetic field — a phenomenon known as paramagnetism. All other common gases (except nitric oxide) have a negative magnetic susceptibility. By measuring this property or changes derived from it, the concentration (or more strictly the partial pressure) of oxygen can be determined.
>
> Three types of paramagnetic instrument commonly exist: magnetic wind instruments, Qwinke analysers and Pauling's dumb-bell cells. The detailed operation of these individual types of sensor is quite complex and can be found in detail in Cooper(1987)[24].

The principal advantage of these instruments is that they are not subject to interference from any other commonly encountered gas, as oxygen is almost the only one which is paramagnetic.

A drawback in the use of these instruments is that most of them actually respond to changes in partial pressure of oxygen and not concentration directly. This means that the reading will fluctuate with changes in atmospheric pressure and, where instruments are calibrated for simplicity to read in percentage oxygen, under some conditions serious error may result. For many applications this is not critical, but for accurate measurements it is important to adjust all partial pressure instruments, however calibrated, to take account of barometric pressure.

Both paramagnetic and electrochemical instruments will normally provide a sufficiently accurate indication of the oxygen concentration for most situations.

10.3.3 Davy lamp

This device has been widely used to indicate oxygen deficiency for many years. It cannot be truly described as an instrument as it has no output meter or digital indication, nor can it give an alarm signal in the accepted sense. The extinguishment of the lamp is usually taken to indicate a hazard and this occurs at around 15.9-16.6% oxygen. Although not expected to have wide use it may still serve to check atmospheres before entry into tunnels and sewers.

10.4 HYDROGEN SULPHIDE, CARBON MONOXIDE AND HYDROGEN

10.4.1 Hydrogen sulphide

Electrochemical cell. This type of detector uses a wide range of detector cells to monitor different gases. Inorganic gases such as hydrogen sulphide are measured using a coulometric cell. The gas reacts within the cell creating a current which is proportional to the amount dissolved. A measurement is then made of the number of coulombs of current produced.

Detector tubes. These consist of a chemical reagent adsorbed on to an inert support within a glass tube. When air is drawn into the tube the concentration of a particular contaminant will be recorded by the length of the stain in the reagent. Selectivity is achieved by using a reagent that only reacts with the particular contaminant to be measured. For hydrogen sulphide the reagent commonly used is lead acetate, which produces a brown stain.

Test paper. Similar to detector tubes, test papers provide another means of on-the-spot analysis. They are impregnated with a reagent specific to the gas of interest. A chemical stain is produced, on exposure to the gas, of an intensity which is proportional to the gas concentration.

These devices are often used as personal monitors.

The above techniques are the most common, but there are also instruments that use the following sensors.

Semiconductor sensor. These devices rely on the adsorption of the hydrogen sulphide gas on to a heated oxide surface. The subsequent reaction on the surface of the oxide produces an electrical conduction change in the oxide which is related to the amount of gas adsorbed.

Photoionisation. In this type of analyser the hydrogen sulphide gas molecules are subjected to an ultraviolet light source which leads to the ionisation of the molecules. The ions are then drawn to a collector electrode which creates a current proportional to the concentration measured.

10.4.2 Carbon monoxide

The most popular detection method is the electrochemical cell. Other techniques used are: semiconductors, detector tubes, thermal conductivity, and infra-red.

10.4.3 Hydrogen

The most popular detector types for hydrogen are catalytic and thermal conductivity devices. Also used are electrochemical cells, semiconductors and detector tubes.

10.5 MULTIPLE GAS DETECTION

10.5.1 Portable analysis kits

As well as single instruments capable of performing multiple-component analysis, it is also possible to purchase portable analysis kits which will enable multiple components to be measured. These kits consist of two or more different types of detector, either as separate entities with individual sampling inputs or as a single sample input which is connected to each detector. A typical combination of detectors would include a catalytic detector for flammable gases, a thermal conductivity detector for toxic inorganic gases and an electrochemical cell device for oxygen determinations.

10.5.2 Portable gas chromatographs

Until recently the use of gas chromatographs has been limited to off-site analysis, but now portable instruments are available, making on-site identification of the components of landfill gas possible. The main features of these instruments are:

- small and of low weight (about 5kg)

- fast run times

- sensitivity (minimum detectable concentration of 1ppm)

- half the price of a conventional laboratory-based machine.

Their main disadvantage is that they are of lower resolving power than their laboratory-based counterparts, being able to identify only 30-40 compounds.

10.6 SUMMARY

Once users have decided on the requirements for gas detection instruments, they can then examine a wide variety of different types and make a choice as to which will suit their needs. Each one will have its own advantages and disadvantages. Some sensor types such as catalytic oxidation and flame ionisation are specific to a particular type of gas (flammable); others, such as electrochemical cells and detector tubes, can be used to detect a number of different gases, e.g. oxygen, hydrogen sulphide and carbon monoxide.

The presence or absence of certain gases may also have an effect on the sensor accuracy, and hence the conditions under which a sensor is to be used must be borne in mind when choosing an instrument utilising a particular sensor. Many devices are now available which attempt to overcome this problem by incorporating two or more different sensors within one instrument. This enables the user to check the concentrations of gases that may affect the readings of the gas of particular interest.

11 Gas sampling points

11.1 Introduction

If reliable representative results are to be obtained from a site investigation attention must be paid to the design of sampling points and sampling methods. It may be very difficult, if not impossible, to obtain a truly representative gas sample, since most methods for sampling inevitably involve some disturbance of the ground. This fact must be recognised and suitable steps taken to minimise the problem whenever practical. Each of the methods of sample point construction described below should be considered as complementary techniques which will be appropriate for use in combination as well as individually, in appropriate circumstances.

11.2 TEMPORARY SAMPLING POINTS

In the initial stages of an investigation it is advantageous to take measurements quickly over a wide area. To achieve this, simple probing of the ground by spiking or the use of temporary shallow probes can quickly yield useful information.

11.2.1 Simple spiking

This is a quick and easy method of taking a gas measurement. A metal spike is forced into the ground and then pulled out. The tube from the gas detection instrument is then placed into the hole and a gas sample drawn into the device for measurement.

The disadvantages of this technique are:

1. Gas measurements will not be highly accurate due to the possibility of air ingress to the hole.

2. Loose earth or water may clog the gas sample line.

3. The depth to which sampling is possible is limited to 1-2m.

11.2.2 Shallow probes

Another simple method of probing for gas in the first stage of an investigation in soils and refuse is to sink a small hollow rigid pipe into the ground, to a depth of about 1m, with a tube connection to a gas detection device (Figures 12 and 13). The probe itself may be hammered directly into the ground or a pilot hole made with a length of steel rod prior to inserting the probe.

One type (Figure 14) found to be particularly good consists of a 1m long 25mm diameter steel tube pointed at one end. The lower half of the tube is perforated at 45mm intervals with 4mm holes countersunk by 2mm, intended to prevent clogging by soil. After inserting the probe the hole is sealed at the surface to minimise the ingress of air, and samples may be withdrawn immediately thereafter from a sampling outlet near the head.

The effectiveness of this technique is very dependent on cover thickness and permeability. Gas concentrations comparable to those from deeper probes can only be measured on sites with thin or permeable covers.

The benefits of shallow probing are:

1. The equipment can be cheaply and easily supplied.

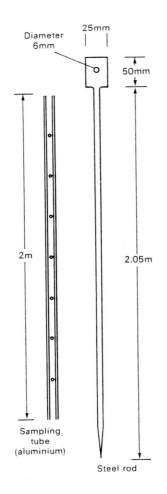

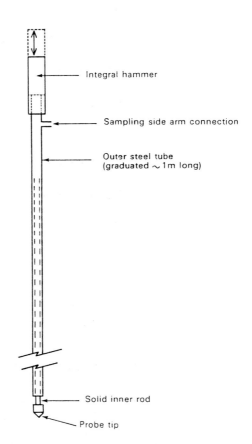

Figure 12 *Diagram of apparatus for sampling landfill gases*

Figure 13 *Soil gas probe (Research Engineers Ltd) representative of the type used by Flower*

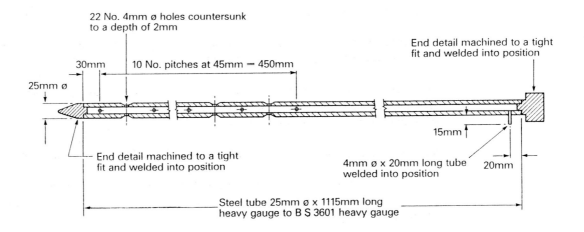

Figure 14 *Shallow spiking probe*

2. It does not require specialist operators.

3. Very little site disturbance is involved.

4. The technique is quick to use and has no major safety implications.

The problems encountered when using this technique are:

1. Air diffusion can reduce the measured gas concentrations.

2. The small sample volume can give misleading results.

3. No visual assessment or sampling of fill materials can be made.

4. Probes are easily blocked by debris and water.

5. There is a danger of over-pumping.

When using narrow bore-tubes there is a possibility of over-pumping which can result in air being drawn in from the surface, thus diluting the gas sample. It is particularly important to eliminate this when sampling tubes are being used to take samples for laboratory analysis.

11.3 PERMANENT AND SEMI-PERMANENT SAMPLING POINTS

For subsequent stages of a site investigation permanent or semi-permanent sampling points will have to be installed. A regular grid is often the best initial arrangement although no hard and fast rules on the size of the grid exist. A large number of sampling points will not necessarily give a clearer picture of the conditions on site, but, interpretation and interpolation of results obtained on too coarse a grid will be difficult. As a starting point, the grid locations should not be greater than 60 m and probably do not need to be closer than 30 m.

Permanent sampling points should not be left open. An open sampling point may:

• allow air into the ground thus reducing locally any anaerobic activity

• allow the release of large amounts toxic and flammable gases

• slow down equilibration of the sampling point with the surrounding ground

• dilute the ground gases, resulting in inaccurate concentration measurements.

In addition, to prevent tampering by vandals, the top of the sampling point should be protected as far as possible with some form of vandal-proof cover.

For ease of sampling it is advisable to have an inlet and an outlet probe permanently fixed at each point. Each should be fitted with an isolating valve which can be used to seal the sampling point between measurements.

Sealing the sampling point will assist its equilibrium with the surrounding ground conditions.

There are a number of techniques available for the construction of permanent sampling points. These should be considered as complementary rather than independent as each will have an application in most site investigations.

11.3.1 Shallow probes

If correctly constructed and protected, simple shallow probes can be used as semi-permanent sampling points for repeated measurements. A simple construction is shown in Figure 15 in which a hole is drilled and the fixed sampling point is fitted.

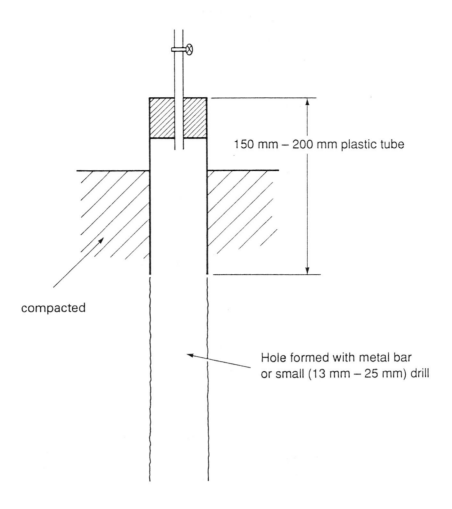

150 mm – 200 mm plastic tube

compacted

Hole formed with metal bar
or small (13 mm – 25 mm) drill

Figure 15 *Shallow probe sampling point*

11.3.2 Deep driven probes

Sampling probes may be installed to depths of about 10m by a compressor-driven hammer impacting on a collar which grips a hollow steel tube approximately 50mm in diameter with a hardened, solid nose-cone (which is subsequently left in the ground on extraction of the main probe). The hammer action drives the probe into the ground a short distance with each blow. Extension tubes can be fitted as necessary to allow the probe to be driven to the required depth.

The hammer may be used to install a perforated probe directly, or the driving probe may be a hollow tube into which a smaller 35mm diameter perforated pipe can be inserted after installation, and pneumatic jacks used to remove the outer steel casing, leaving the monitoring pipe in place (Figure 16). On extraction of the casing, surrounding soil and wastes are allowed to collapse onto the pipe and a protective section with a threaded cap is then attached which can be concreted in position for security.

As well as hammer-driven probes, a new technique has recently been developed from apparatus used for reinforcing earth slopes, i.e. soil nailing, where explosively driven probes are used. Whilst this allows probes to be driven deep into the ground, it requires the use of specialist equipment and operators.

A stabilisation period for the sampling point of up to three weeks may be required for some installations, although reliable data can sometimes be obtained after only a few days, as there is usually relatively little ground disturbance.

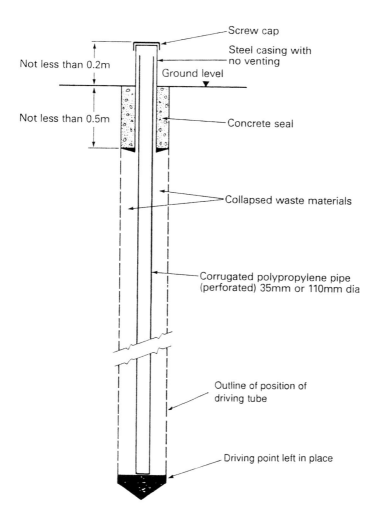

Figure 16 *Gas probe left in place after withdrawal of driving tube*

The advantages of this technique are:

1. There is minimal risk if the site is highly toxic or on fire.

   Where an underground fire is suspected the site may be unable to support heavy equipment. If toxic materials are known to be present, disturbance of the material and extraction to the surface should be avoided where possible.

2. Installation is rapid.

3. It can be used in small areas or where access is restricted, such as gardens.

   This technique may be particularly useful in confined spaces or gardens where access for the larger plant required for other types of sampling point may be difficult, and where minimal disturbance of the site is desirable.

The main disadvantages of this technique are:

1. Specialist plant and operators are required.

2. Deeper probes may deviate from the vertical during the driving process and thus the exact location of the measurement will be uncertain.

3. Installation may be prevented by obstructions, e.g. hard-core or timber.

4. Equipment mobilisation costs may be high depending on the number of probes installed.

5. There is no visual indication of the groundmass penetrated, although the ease or difficulty of driving may provide some indication of the materials.

11.4 BOREHOLES

A permanent or longer-term monitoring point will usually need a hole drilled into the ground. This may be lined with a perforated pipe made of a material that will not adsorb the gases (uPVC is sometimes used), or left unlined and sampling probes installed directly.

Designs for the arrangement of probes within the hole vary, but generally the gas probe is inserted to the required depth and void spaces are backfilled with pea gravel or similar open-pored material. The top of the hole is sealed to prevent ingress of air and to protect personnel from any toxic gases emanating from the borehole. A number of different arrangements are illustrated in Figures 17-21.

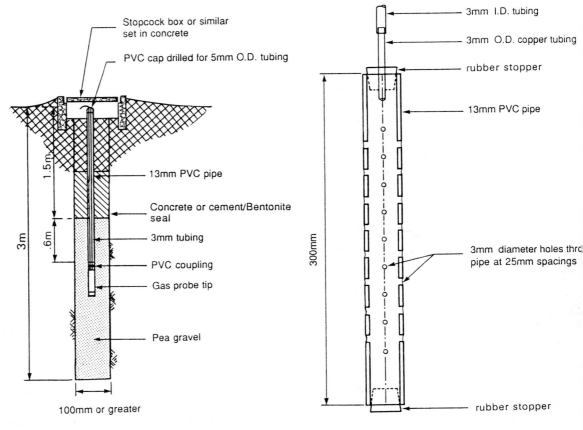

Figure 17 *Typical borehole installation*

Figure 18 *Typical gas probe tip in borehole installation*

The perforated portion of a probe or lined section of a hole may extend down the full depth of the hole, apart from say 1 m from the surface, or may be a short section at the base of the hole. A number of samples should be withdrawn from different depths inside the hole to account for any variations in concentration that might occur with depth.

As an alternative to using a single probe to sample from different depths within a borehole, more than one probe[28,31] may be installed (see Figure 19) at different depths in the same hole. Multiple installations on a single stem using packers or grout seals[29], such as shown in Figure 20, provide a practical method which avoids the difficulties of depth control and backfilling around two or more pipes. In these arrangements lengths of the borehole between the zones being monitored must be sealed to prevent gas migration from one zone to another. These sealed positions should be chosen in relation to the surrounding material. The advantage of multi-zone monitoring is that it can perhaps give more reliable information about the activity of a site at different depths.

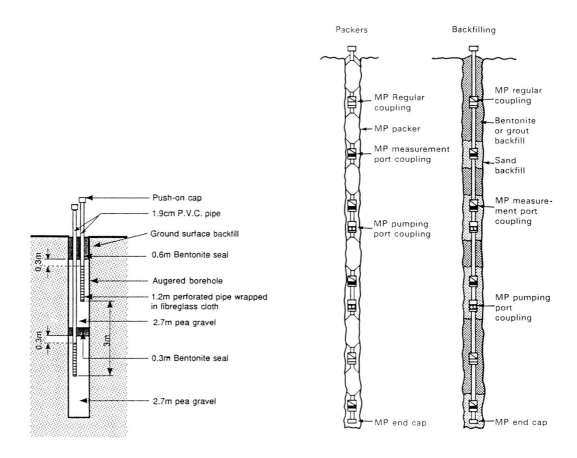

Figure 19 *Typical nested installation in one borehole*

Figure 20 *Multi-probe installations*

The other main advantages of any of the borehole types of sampling point are:

1. The depth of the sampling point is not limited by the construction method.

2. Boreholes also provide hydrogeological and geotechnical information about a site.

They are, therefore, multipurpose tools, useful in most site investigations.

The disadvantages of borehole construction are:

1. It disturbs the soil.

   This is a feature common to other types of sampling point and not a significant disadvantage. There is also some advantage since material drawn to the surface will give an indication of the underlying ground conditions.

2. Drilling, particularly through fill material, may be difficult and expensive.

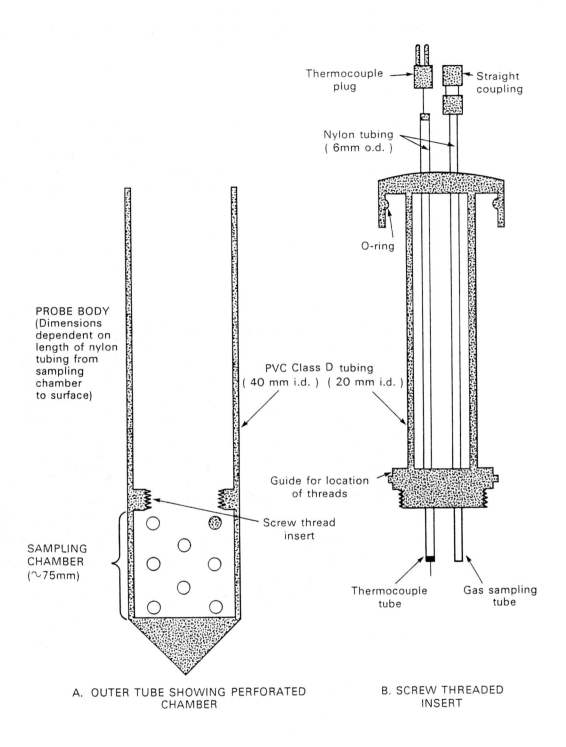

Thermocouple plug

Straight coupling

Nylon tubing
(6mm o.d.)

O-ring

PROBE BODY
(Dimensions
dependent on
length of nylon
tubing from
sampling
chamber
to surface)

PVC Class D tubing
(40 mm i.d.) (20 mm i.d.)

Screw thread
insert

Guide for location
of threads

SAMPLING
CHAMBER
(~75mm)

Thermocouple
tube

Gas sampling
tube

A. OUTER TUBE SHOWING PERFORATED
CHAMBER

B. SCREW THREADED
INSERT

Figure 21 *Gas sampling probe*

3. It brings contaminated material to the surface.

   On sites known to contain toxic wastes the amount of material brought to the surface should be minimised whenever possible and precautions taken for its handling and disposal.

Following the construction of a borehole sampling point, because the ground is disturbed and possibly becomes slightly aerated and/or compacted, the ground should be left to settle before measurements can be considered representative. Opinion varies over exactly how long this should be, and much will depend on the degree of disturbance and other conditions on the site. The period may be a few days to several weeks, but there are no established recommendations.

11.5 TRIAL PITS

Another method of sample point construction is to dig pits with a mechanical excavator. Perforated standpipes are then installed at the corners of the pit, backfilled (usually with the original material) and left for several days before taking measurements. With this method depths to 5m can be seen (6m with larger excavators) quite easily, although care has to be taken about the stability of the sides. A typical example is illustrated in Figure 22.

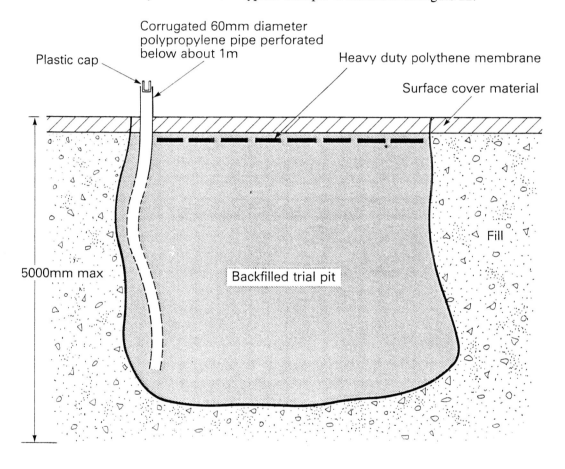

Figure 22 *Trial pit gas probe within fill*

The advantages of using trial pits are:

1. They are, in most cases, cheaper to install than boreholes.

2. They provide visual indication of the groundmass.

> *The technique enables the operator to get an indication of the ground conditions for the area excavated, from which the degree of degradation can be ascertained. For reasons outlined earlier, digging trial pits on sites known to be contaminated with toxic wastes should be minimised, and then only if proper precautions are taken.*

3. They can usually be excavated and backfilled quickly.

Trial pits are, therefore, particularly useful in a preliminary investigation of the site for detecting gas and other types of contamination.

Provided adequate safety precautions are taken, a crude method of determining the presence of gas before backfilling is to cover the trial pit overnight with a plastic sheet or tarpaulin. Samples of the atmosphere under the sheet can then be taken to find out if flammable or harmful gases are present.

Precautions must be taken to ensure that personnel do not enter a pit. Staff should be forbidden to do so. In addition, as a general rule, personal monitoring devices should be worn by people working on the pits. Similarly, the potential for an ignition to occur within a pit should be minimised by the use of 'safe' equipment and, where necessary, the removal of smoking material from working personnel.

The principal disadvantages of the trial pit as a sampling point are:

1. The limitation of the depth that can be examined.

2. The considerable disturbance of the soil created by digging the pit.

3. The possible formation of air pockets on backfilling.

4. Odours from the pit may create a public nuisance and environmental health hazard.

Because of the substantial soil disturbance, a longer period will be needed before ground conditions return to normal and measurements can be made. Both these factors may place some limitations on the use of trial pits. However, the ease of construction, the additional benefit of a visual examination of sub-surface conditions, the possible benefit of allowing a route for the gas to escape, ease of detection and the cost advantages make trial pits a satisfactory addition to boreholes. This is particularly so for the initial stages of an investigation or on sites with a relatively shallow depth of fill.

11.6 SURFACE SAMPLING APPARATUS

As an alternative to the use of sub-surface sampling methods, although probably not widely used, attempts have been made to characterise gas emission from collection points situated on the surface. The procedure is to invert a large open container over a small area of the site so that the gas emanating from the surface can be collected inside. Measurements are obtained either by grab sampling or continuous monitoring of the internal atmosphere with a suitable instrument. The container, typically a plastic water tank or steel drum, is sealed at the ground surface with a material such as bentonite and should have some form of pressure relief system to avoid pressurising the container. An example of this type of apparatus is shown in Figure 23.

11.7 SUMMARY

Various designs of sub-surface sampling points are in use, ranging from the simple spiking of the ground to form a hole from which a gas sample is taken, to the use of more expensive boreholes and trial pits. Each method has its own limitations many of which can be overcome by using a combination of techniques. The choice of sampling point will, to a large extent, be determined by the site conditions and will also depend on whether information is required about the sub-surface materials and groundwater.

Surface sampling of gases can be undertaken by either collecting the gas inside a container such as a plastic water tank or steel drum, or by taking direct measurements using a flame-ionisation detector. Surface sampling, if used at all, should not be undertaken without sub-surface sampling.

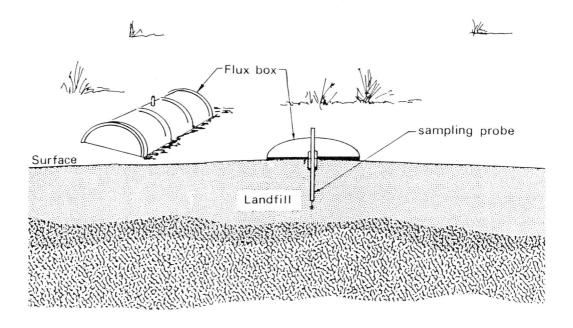

Figure 23 *Surface sampling*

12 Sample collection

12.1 INTRODUCTION

Obtaining a representative sample of gas from a sampling point is as important as the instrumentation used and the sampling point construction. The procedures for determining composition and concentration of gas samples are relatively easy providing simple precautions are taken and the limitations of the measuring instrument are understood.

12.2 GROUNDWATER LEVEL

Before any sampling is attempted it is advisable to determine the water table level, so that sample lines/probes are not placed too deeply resulting in water being sucked into the instruments or collection vessels. It may often be worthwhile installing a simple water trap (Figure 24) to prevent water accidentally getting into an analyser. This does however have the disadvantage of increasing the sampling system volume.

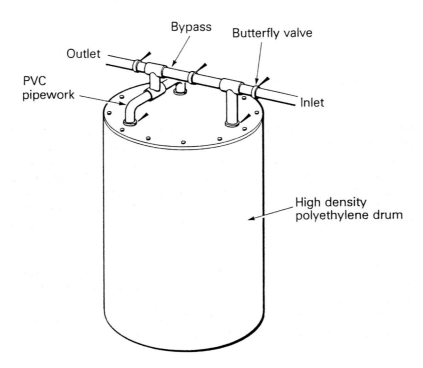

Figure 24 *Gas dewatering drum*

12.3 SAMPLE SYSTEM − ON SITE MEASUREMENT

On site, the simplest procedure is to connect the measuring instrument to the sample point and withdraw samples until a steady reading is obtained. To minimise the influence that withdrawing the sample from the sampling point has on the measurement itself, ideally the volume from which the sample is being taken should be as large as possible, while the volume of the sampling system (probes, sampling lines and other external apparatus) should be as small as possible. Sampling lines should not be so narrow, however, that they restrict the gas flow.

If determining more than one component with different instruments this procedure will have to be repeated for each gas. Alternatively, the analysers may be connected to the sample point in parallel (Figure 25) and measurements made simultaneously. It is advisable to fit non-return valves on the inlet to each analyser to prevent air being drawn back through them by the other instruments. If an external pump is available, care should be taken to ensure that the pump neither draws in air nor over-pressurises the analysing equipment.

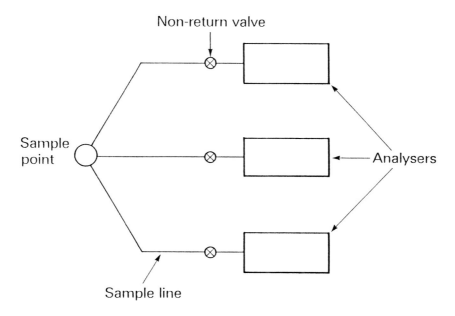

Figure 25 *Schematic of parallel sampling system*

Normally the analysed gas sample would be vented into the atmosphere, but an alternative procedure is to recirculate it back into the borehole or standpipe. Providing the volume of the sample lines is small relative to that of the sampling point, the steady state reading obtained will be close to the equilibrated concentration in the ground. The off-take and reinjection points should be as far apart as possible to ensure maximum circulation within the sampling point (Figure 26). This method has some advantages in that it reduces the possibility of air being drawn into the ground as a result of the sampling procedure, but it cannot be employed when using instruments that destroy the sample in measurement.

It may often be the case that the amount of sample available for measurement will be physically limited by the dimensions of the sampling point and other factors such as high water table levels and low rates of replenishment. Under these circumstances reasonable readings may be obtained if the recirculation method is used. Alternatively an *in-situ* sensor may be used, but again this should be of a type which does not destroy the sample.

12.4 SAMPLE COLLECTION FOR OFF-SITE ANALYSIS

The removal of samples for off-site analysis can be done in a variety of ways. The choice of method is to some extent arbitrary, but some types of subsequent analysis, particularly if aimed at determining the minor components in a mixture, are facilitated by certain special methods.

It must be remembered that when samples are collected for analysis a reduction in the concentration of gas on the site may result. Hence it is advisable to collect as small a sample as possible, especially on sites emitting small volumes of gas.

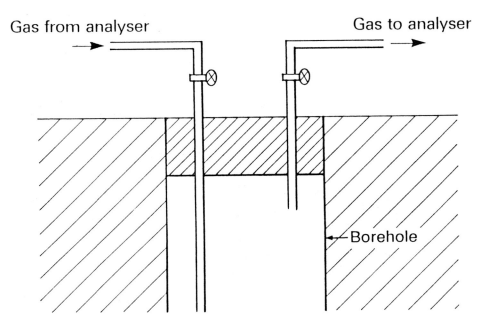

Gas from analyser   Gas to analyser

Borehole

Figure 26 *Recirculation of gas from a sampling point*

A simple and widely used method of collecting gas samples for analysis is to use pressurised sampling cylinders (Gresham[R] Tubes see Figure 27). These are designed to be used with a hand pump to compress the gas sample into a small cylinder made from either aluminium alloy or preferably stainless steel. Various types of cylinder are available: 'flow through' (i.e. with a valve at each end of the tube) to allow pre-purging with the sample gas, or 'closed end' which should be filled and emptied several times to ensure a representative sample is obtained. They are available in various sizes from 15 ml up to 110 ml.

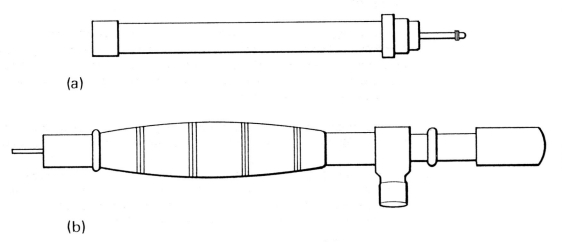

(a)

(b)

Figure 27 *(a) Gas sampling cylinder (single headed)*
(b) Gas sampling pump (two stage)

Another simple arrangement (Figure 28) for obtaining a sample consists of a gas sampling vessel, which may be sealed at both ends by taps or valves. The vessel is connected to the sample point in series with a vacuum pump or hand aspirator to provide suction. A by-pass line may also be fitted, if desired, to allow monitoring of the sample to proceed while the sampling vessel is being removed. Gas is withdrawn by the pump through the sampling vessel until at least three to five volume changes of the vessel have been carried out. If, as is advisable, some continuous monitoring is being performed, the vessel should be flushed until a steady reading is obtained at its outlet. Having filled the vessel, the taps are closed, and it is removed.

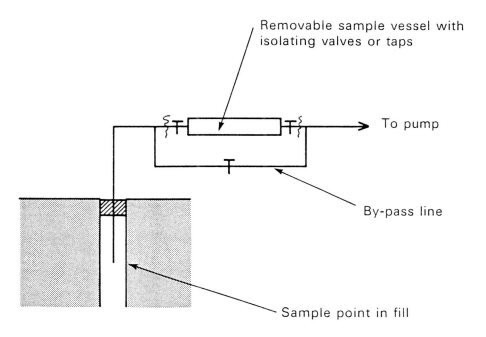

Figure 28 *Principle of a simple collection system*

It is important that for both these methods the sample should be clearly labelled. Labelling should include: date, time, location and, whenever possible, the approximate concentration of at least one major component.

It may also be advisable with these methods to flush the sample tube with an inert gas such as argon before use. This will ensure that, if the tube is not thoroughly flushed with sample gases during the sampling process, their true concentration can be calculated by determining the amount of residual argon.

The sampling vessel can be pre-filled with a liquid as an alternative method of providing suction to withdraw the sample if no a pump is available. Emptying the liquid from the vessel under gravity provides suction to fill it from the sampling point. The liquid used should be acidified distilled water, with added methyl red indicator, since the low pH minimises the dissolving of carbon dioxide in water.

A comparison of these two methods of filling a sampling vessel showed the sampling technique had no effect on the results, although the pumping system was found to be easier to use[30].

A further alternative to the straightforward sampling vessel is to use a pre-evacuated vessel into which a sample can be withdrawn directly from the sampling line. A possible advantage of this technique is that it reduces the total volume of gas removed from the site, thereby minimising the potential for air ingress due to sampling procedure.

12.5 MATERIALS

12.5.1 Tubing

It has already been mentioned that the volume of sampling lines should be kept to a minimum and short narrow lengths of tubing used wherever possible. Care must be taken to ensure that the tubing is not so narrow as to restrict the gas flow or become easily clogged by water or debris which may accidentally be withdrawn with the gas.

The tubing used needs to be robust and flexible, although the ideal material for fixed parts of an installation is internally clean stainless steel. For the determination of the major gas components the choice of materials is probably not critical since other factors are likely to have a much more significant effect on the measurements made. Any general purpose flexible tube may be used although the best materials are probably neoprene, polypropylene or Tygon[R].

For the measurement of low concentrations of gas and trace components the adsorption and desorption properties of the tubing can be important. If collecting samples for detailed laboratory analysis the use of an inert flexible tubing such as Viton[R] is preferable.

12.5.2 Drying agents

The use of drying agents to remove water vapour from the gas sample is desirable to minimise the risk of damaging the equipment and to ensure reliable measurement if using instruments that are sensitive to moisture.

Large amounts of water and other condensates should be collected in a water trap, but drying agents will be required to remove residual amounts of water vapour. Care over the choice of agent should be taken since some agents, particularly silica gel, will absorb gases such as carbon dioxide. This problem is exacerbated as the agent becomes wet. For some infra-red analysers the recommended drying agents for other than the most sensitive analysis are either anhydrous calcium chloride or anhydrous calcium sulphate. For sensitive analysis magnesium perchlorate is probably the most suitable drying agent.

12.5.3 Sampling vessels for sample collection

A sample collection vessel may be made of glass although for site investigations metal vessels would prove more durable. Ideally, these should be made from stainless steel, though brass or monel may be suitable for most purposes where the gas is not corrosive.

12.6 SAMPLING BY ADSORBENT MEDIA

The methods described above for sample collection are generally applicable, but where detailed analysis of minor components is required more specialised techniques may have to be employed. These utilise adsorbent columns or cold traps which selectively trap, and therefore concentrate, particular groups of compounds. These are specialised techniques which require a relatively high degree of expertise.

Sample tubes made of glass and containing a high-activity sorbent can be used to collect gases from air samples. A sample is collected by breaking off the end tips of the tube and drawing the air/gas sample through. The chemical is adsorbed onto the sorbent in the tube and push-on caps fitted to seal the ends. It can then be taken to a laboratory where the trapped chemicals are solvent-extracted or thermally desorbed before being analysed by gas chromatography.

This method of sample collection is not widely used to collect gas samples emanating from the ground.

12.7 SUMMARY

There are basically two ways of approaching the task of collecting and analysing a gas sample from a gassing site. One method involves withdrawing a sample from a sampling point directly into a measuring instrument, to give an on-the-spot reading. Alternatively, a sample can be collected and taken to a laboratory for a more detailed analysis. Various methods of sample collection can be used; the most common involve pumping the gas into a cylinder or vessel. A less popular technique uses adsorbent media to collect the gas sample, which is then desorbed in the laboratory.

Care should be taken when choosing the materials to be used for the tubing and drying agents, as adsorption of some of the constituent gases in the sample may occur.

13 Meteorological data

13.1 METEOROLOGICAL EFFECTS

It is important during the monitoring of a site for gas emission that the climatic conditions prevailing before and at the time are recorded. The information obtained from these records may help in the interpretation of the gas monitoring data and explain any anomalies. Measurements should be taken on-site and backed-up by data from the Meteorological Office.

The four parameters generally recorded are:

- temperature

- atmospheric pressure

- rainfall

- wind speed/direction.

13.1.1 Temperature

No clear relationships have been found to exist between temperature variations and gas concentrations or gas emission rates, though in shallow landfill sites high air temperatures may encourage increased landfill gas generation, the optimum temperature range being 35° to 45°C. If the ground becomes frozen the surface of the site will effectively be sealed, inhibiting gas evolution, but causing the gas to increase its lateral migration.

13.1.2 Atmospheric pressure/rainfall

Gas concentrations do not appear to be influenced at all by variations in atmospheric pressure/rainfall. Gas flow rates, however, can be affected. On some sites a fall in atmospheric pressure can lead to a higher gas emission rate from the ground, although this relationship is not totally clear and depends on whether the area of influence extends over the site and surrounding strata, or just the surrounding strata. Another variable that determines the rate of emission of gas from a site is the rate of change of atmospheric pressure: a rapid fall will increase the rate of emission. Not all gas flow rates are affected in this way, the effects of atmospheric pressure/rainfall are generally site specific.

Over periods of prolonged dry weather, site-covering material, especially if formed of clay, will tend to crack, causing increased gas emission from the site. In wet weather rainfall percolating downwards can close up gas pathways.

Gas migration can also be affected by barometric pressure and rainfall in the following ways:

- by creating pressure gradients

  Low atmospheric pressure creates a favourable pressure differential for the gas to migrate from its source to the surface.

- by increasing lateral migration

  Rainfall tends to seal the surface of the ground, temporarily encouraging the gas to migrate further than it otherwise might.

- by increasing gas emission.

  Rainfall can alter the water table. A rising water table can pressurise the gas and force it out, possibly cutting off previous migration routes. Conversely, a rapidly falling water table may draw air into the ground.

13.1.3 Other effects

On sites such as in-filled docks there are significant tidal effects which alter the gas emission patterns. When the tide is in the water table will rise, forcing the gas out at a greater rate than when the tide is out. Wind speed and direction can make gas emission readings difficult to take, particularly when using hot-wire anemometers.

13.2 SUMMARY

Climatic influences on the gas regime on a site should not be underestimated when taking gas measurements. Changes in temperature, atmospheric pressure, rainfall, wind speed/direction and tidal effects can all influence the gas readings. It is therefore important that these parameters are recorded along with the gas measurements, and their effects on the gas noted if the collected data are to be interpreted correctly.

14 Methane in groundwater

14.1 INTRODUCTION

Methane which is dissolved in groundwater under pressure may be released into the atmosphere when pressure is relieved. This can occur if the water reaches the surface, or if the groundwater is penetrated by a low-pressure construction such as a pipeline, tunnel or mine. Such gas releases may represent a hazard directly by entering the construction, or indirectly as a result of water or gas tracking along the direction of the tunnel and being released where it breaks the surface. It is therefore important, when involved in underground construction, that the groundwater is assessed for dissolved methane and other gases.

14.2 SAMPLE COLLECTION

In order to collect good groundwater samples for dissolved gas analysis it is important that the design of the sampler should meet certain criteria[34].

It should not allow sample concentration changes, which could arise from:

- changes in temperature or pressure

- leaching from the sampler

- contamination or sample loss during handling.

The sampler should also have the following attributes:

- be able to collect samples reproducibly

- be easy to use

- be inexpensive

- be used in a variety of wells/boreholes at varying depths.

For dissolved gases such as methane, the presence of a headspace volume in the sampling apparatus can lead to the degassing of the methane into the headspace volume. Hence, analysis of the water sample will give low concentrations. Thus the sampling apparatus should ideally have no headspace (if total degassing occurs analysis of the resulting headspace gas by gas chromatography can be used to determine the gas concentration).

Various types of sampler[35,36] have been developed which can be used either as 'dedicated' samplers (permanently installed devices) or as 'single-event' samplers (used to collect one sample per test); these include:

- grab samplers − bailers, syringes

    Bailers are cheap, portable and simple to use. The bailer consists of a tube with a ball type seal at one end, with the other end open. The instrument is lowered down the well ball-end first until it reaches the required depth within the groundwater. A sample is collected by lowering and raising the bailer which opens and closes the ball seal. Bailers are usually made of Teflon(R), PVC or stainless steel.

    Syringes are relatively inexpensive, portable, easy to clean and can be made from inert materials. Their main disadvantage is that they are not very good for collecting large volumes of samples.

- positive displacement devices

These devices, such as centrifugal pumps, submersible piston pumps and gas squeeze pumps, are used when degassing of the contaminants is a prime consideration. Their main disadvantages are that they are quite expensive, difficult to clean and need well diameters of at least 50mm.

- suction-lift devices

Centrifugal and peristaltic pumps are usually employed as they are inexpensive, portable, easy to clean and simple to operate. However, they can only sample down to 6−7 m and they may change the chemical composition of the water.

The above devices will show significant bias and poor precision unless great care is taken in sample collection and transfer. Grab sampling devices are also very dependent on the expertise of the user and the field conditions[35].

- sorbent samplers[34]

Groundwater is drawn through a cartridge of a sorbent material which collects the dissolved material. The sample is then taken to a laboratory where the adsorbed material is removed by either solvent extraction or thermal desorption.

The advantages of this method are: low detection limits, ease of transport/storage, and that samples can be collected from small diameter (10 mm) wells. The disadvantages are that the dissolved gases are not very well retained by the sorbent materials and sampling of very contaminated groundwater can cause overloading of the sorbent and analytical equipment.

- reservoir samplers[34] (Figure 29)

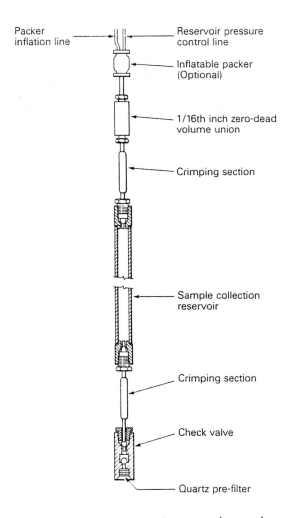

Packer inflation line — Reservoir pressure control line

Inflatable packer (Optional)

1/16th inch zero-dead volume union

Crimping section

Sample collection reservoir

Crimping section

Check valve

Quartz pre-filter

Figure 29 *Components of a reservoir sampler*

vane' and the 'hot wire'. The hot wire is more sensitive and therefore most commonly used for the measurement of low velocity flows.

Hot wire anemometer. The hot wire anemometer operates on the principle of change in thermal conductivity caused by a gas flowing over a heated wire (usually platinum). As the velocity increases, so the effective thermal conductivity of the gas increases (the hot wire is cooled), producing a change in resistance which can be used to measure the flow velocity.

Spinning vane anemometer. This device operates by counting the number of revolutions per unit-time of a vane which rotates in the presence of a gas flow. The anemometer must be calibrated in a known flow of gas.

Pitot tube. The Pitot tube measures velocity by determining a pressure difference between static and dynamic pressures inside the device when placed in a flowing gas stream. The sensitivity of measurement is determined by the minimum pressure difference that can be measured.

When using any of these devices the user should be aware that a number of factors may cause error in the absolute accuracy of the measurements being made. For precise measurements these have to be taken into account. They are:

1. The effect of changes in composition of the gas stream (the instruments are most commonly calibrated in air).

   Flow measurements can be affected by the chemical composition of the gas mixture being measured which can change its thermal conductivity and density.

   On landfill sites the gas usually encountered consists of a range of methane/carbon dioxide/air mixtures. Due to the differences in the thermal conductivity values for these three gases, considerable variations can occur in the hot wire anemometer performance depending on the composition of the gas mixture.

   If the anemometer is calibrated for air flows, large amounts of methane (which has a higher thermal conductivity than air), can cause the anemometer to overestimate the true velocity. Large amounts of carbon dioxide, which has a lower thermal conductivity than air, can cause an underestimation of the true velocity. The true velocity will only be measured for mixtures with the same thermal conductivity as air. Moisture content and temperature of the gas will also influence the thermal conductivity of the gas mixture and thus affect the readings obtained.

   For similar reasons any change in composition can affect the density of the gas mixture. This is important when using devices such as rotameters which are dependent on this property of a gas for their measurement.

2. The effect on the emission velocity of introducing the sensor into the gas sampling point.

   The presence of the velocity measuring sensor (e.g. hot wire anemometer or Pitot tube) within the gas stream can lead to a significant error in the flow velocity in small diameter pipes (50mm or less). This is because of the proportion of the total cross-sectional area of the pipe occupied by the sensor which reduces the effective volume of the pipe at the measuring point and increases the apparent flow velocity. The size of sensing probe relative to the diameter of pipe into which it is inserted should therefore be kept as small as possible. Recommended values for Pitot tubes are one-fiftieth of the sampling pipe diameter.

3. External effects, e.g. temperature, pressure and crosswinds.

   All flow measurements are affected by variations in temperature and pressure. For precise and accurate measurements these must be taken into account.

   When taking measurements on site the disturbance caused by wind effects can significantly affect the readings obtained. To minimise the effects on measurement caused by crosswinds, a rigid box may be used as a shield. Another method is to fit the top of the sample point with an extension pipe so that the measuring point is effectively at least 0.75m from the outlet.

4. Sensitive anemometers, if hand held, can give readings of 0.03-0.05 m/s when there is no gas flow, simply because of hand vibration.

15.4 INDIRECT METHODS

As well as the direct measurement of flow described above, indirect measurement based on changes in concentration with time can be used to estimate the rate of emission.

15.4.1 Recirculation

This method should only be used if the emission rate is very low or there is no flow or positive pressure measurable using the direct techniques. The procedure involves flushing a sample point or standpipe with nitrogen, air, or an inert gas such as helium (nitrogen is not recommended as it makes the interpretation of the results difficult), to displace all traces of the contaminant to be measured, usually methane. Gas from the sample point is then cycled through the measuring instrument and the change in concentration with time monitored, preferably continuously.

(Note: Instruments which change the composition of the sampled gas cannot be employed for this method.)

The parameter actually being measured is the rate of equilibration of the gas in the sample point with the surroundings due to diffusion and convection; thus the rate of change in concentration will decrease with time. The rate of change must, therefore, be extrapolated to time zero to give the maximum value. From this value and the volume of the sample point a reasonable estimate of the free flow emission rate (typically cm^3/min or hr) can be calculated. The depth of the water (if present) in the sample point must be measured to determine accurately the sample point volume.

The use of a recirculation technique for measuring emission rates is obviously limited to the type of sample points and standpipes in which recirculation is possible. The volume of the sample point or standpipe must also be known. If it is assumed that this is just the volume of the perforated liner then this can be quite easily determined. However, if void spaces exist in the ground adjacent to the sample point these may appreciably increase the volume of the sample point and give rise to a significant error in the measurement.

15.4.2 Flux box measurement

For this technique a box or container is placed over and sealed to the surface of the site where gas is emitting (Figure 23). The box is continuously purged with a known low flow rate of clean air, so as not to overpressurise the box and cause leakage. Gas emitting from the ground also enters the box and is therefore combined with the input gas flow. By determining the steady state concentration of, for example, methane or any other gas exiting from the box, the flow rate of that gas from the ground can be calculated. It is important that the seal between the ground and the box is as good as possible to prevent excessive leakage reducing the accuracy of the results.

15.5 SUMMARY

Measurement of the gas emission rate from a site is a far more difficult task than determining gas concentrations. There are basically two approaches to this problem. Direct measurement techniques can be used, utilising instruments such as rotameters, bubble-meters, hot-wire anemometers, spinning vane anemometers and Pitot tubes. The other method uses an indirect approach based on changes in gas concentration with time. Two systems usually employed are recirculation, which is the best method for low emission rates, and flux box measurements.

16 Continuous monitoring systems

16.1 INTRODUCTION

An automatic data logging system provides the most efficient and convenient method of obtaining and storing data from gas sensors. Some data loggers allow extra memory to be added and other parameters to be measured, such as water levels, conductivity, or pH. Loggers can also be used to trigger alarms when a preset safety level has been exceeded.

There are three basic types of data logging system available:

- central sequential
- local sequential
- in-place sensor.

16.2 Central sequential

This type of system consists of a gas analyser and a data logger which are connected to gas sampling points, e.g. boreholes around the site (Figure 32). The gas analyser sucks the gas from each borehole sequentially through tubes of about 6mm in diameter.

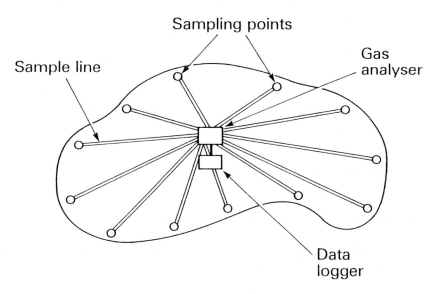

Figure 32 *Central sequential sampling scheme*

There are a number of drawbacks with this system. These are:

1. Any water vapour present in the gas can condense and block the sampling lines.

2. The condensate itself may give off gas, affecting the readings.

   To overcome the problems of condensation, filters are used.

3. The gas has to be flushed right through the sampling lines in order to facilitate an accurate reading.

4. Sampling time may be long if the sample lines are long.

5. A large amount of gas will be withdrawn if the sample lines are very long.

   It is often the case that on many sites the volume of gas in the boreholes will be quite small. Hence this system of monitoring would be inappropriate.

6. Withdrawing large volumes of gas creates a lot of disturbance and causes sample dilution.

7. If a problem occurs with the central analyser/logger all the readings will be lost.

16.3 LOCAL SEQUENTIAL

This system is essentially a localised version of the central sequential system. A smaller number of boreholes are connected to an analyser/logger, which is in turn connected to others in a loop (Figure 33). A master logger can be connected to the system to get data from the local loggers around the site.

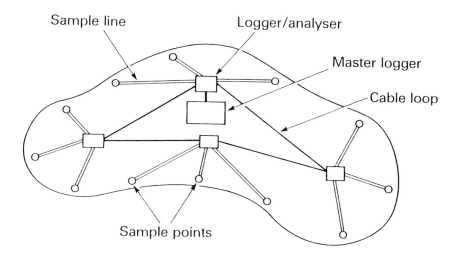

Figure 33 *Local sequential sampling scheme*

The advantages of this system over the previous one are:

1. A problem with one sensor will not affect the readings from the whole site, only those on the loop it is connected to.

2. The nearer you move the data logger to the sampling points, the more accurate the readings become.

3. Sample disturbance, though still present, is reduced.

With both the local and central sequential systems the loop can be expanded or contracted as necessary to include the required number of sampling points.

16.4 IN-PLACE SENSOR

This type of system is designed to record the data obtained from a single sensor. The sensor is placed inside a borehole, with the logger situated on the surface (Figure 34). Each data logger can support up to three sensors and can also be connected to a master data logger if more comprehensive monitoring is required.

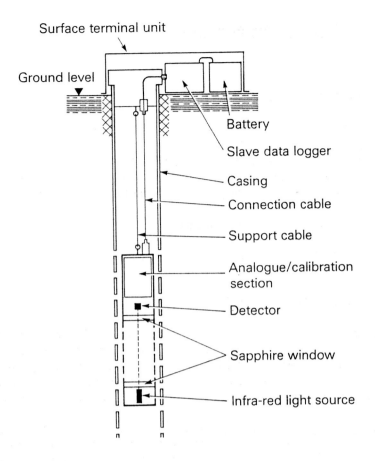

Figure 34 *In-place sensor*

The advantages of this system are:

1. Instantaneous gas readings are obtained.

2. There is no sample disturbance.

3. Individual sensors can be programmed to take readings automatically at preset intervals.

16.5 SUMMARY

Three types of continuous monitoring system are available: central sequential, local sequential and in-place sensor. The central sequential system analyses and logs the data from each sampling point sequentially. The local sequential system does the same but fewer boreholes are connected to the logger, which is also connected to other loggers and a master logger. The in-place system is different entirely as the logger is connected to a sensor placed in the borehole which continually monitors the gas concentration, thus eliminating the need to remove the gas sample from the sampling point.

17 Monitoring in buildings

17.1 INTRODUCTION

Monitoring the concentration of gas within a building is an important part of any post development programme. As with the external environment, the purpose of an internal monitoring system is to ensure that the environment remains safe, and that measures incorporated into the building to control gas ingress are working efficiently. There are two principal considerations in the design of any system for this purpose:

1. What type of detectors are most suitable

2. Where should the chosen detectors be positioned?

17.2 DETECTION SYSTEMS

Most of the detectors and sensors described in previous sections can be used as the basis for a multi-point fixed sampling system, although the most commonly used are probably the catalytic and infra-red based systems. These can be linked to automatic logging and warning systems connected either optically or electronically to a single central microprocessor based controller. The advantages and limitations of various detector types are discussed in Sections 9 and 10.

Whatever system is used there will be an on-going requirement for maintenance, checking and calibration of the sensors. Many systems which are microprocessor controlled have self-check facilities which will assist this process and give an early warning of detector failure. It may also be advisable in sensitive areas to fit two types of detector at the same location: this can help avoid problems of 'common mode failure', i.e. similar detectors failing by the same fault whether electrical or a result of the sampling conditions.

17.3 SENSOR LOCATION

There are many uncertainties and misconceptions concerning the positioning of detectors in buildings to measure gases which may enter from the ground. These tend to arise for two main reasons:

1. The major hazard is often perceived to arise from one gas only, usually methane; therefore the behaviour of the gas is assumed to emulate that of pure methane or natural gas.

2. No account is taken of the influence of other gases (e.g. carbon dioxide) on the nature of the mixture in terms of flammability, buoyancy and toxicity.

For example, landfill gas, as a mixture principally of methane and carbon dioxide, has a buoyancy that varies depending on the exact composition and temperature of the mixture. For a typical mixture, i.e. between 45 and 65% methane under standard conditions, the buoyancy is close to that of air (Figure 35), being either slightly more (65% methane) or slightly less (45% methane) buoyant than air. In addition, if the landfill gas contains nitrogen or is diluted by air before it enters the building, the buoyancy may be very similar to that of air. Under either circumstance the effect of buoyancy may be small and the behaviour of the gas will be governed to a great extent by the natural air flows and ventilation within the building.

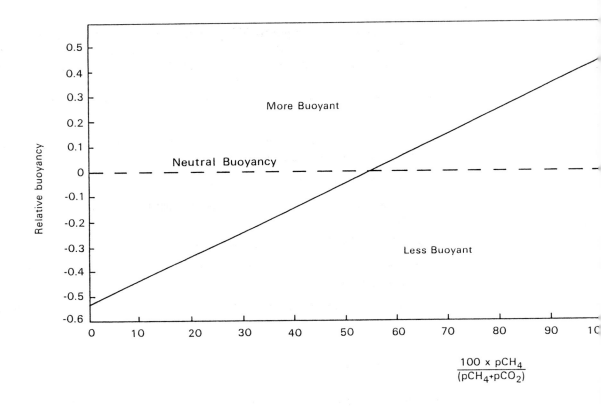

Figure 35 *Buoyancy of methane/carbon dioxide mixtures relative to air*

KEY TO INGRESS ROUTES

1. Through cracks in solid floors
2. Through construction joints
3. Through cracks in walls below ground level
4. Through gaps in suspended concrete or timber floor
5. Through gaps around service pipes
6. Through cavities in walls

Figure 36 *Gas entry into buildings*

In all circumstances where gas could ingress from the ground below a building, sensors should be sited as close as possible to the likely routes of ingress at whatever height they exist. Additional sensors at ceiling height may provide further safety, but they should not be relied upon as the only location for gas detectors. Figure 36 highlights some of the potential routes of ingress for a typical domestic dwelling; corresponding locations on commercial properties would also have to be considered.

Within a building that may be subject to gas ingress from whatever source, suitable gas detectors should be considered in the following areas:

- rooms with little or no natural ventilation

- rooms containing potential sources of ignition, particularly electrical equipment, gas burners or boilers

- underfloor and wall cavities where gas may collect and migrate

- ducts and, if placed within the ground, sewers and drainage channels, which may form routes for gas ingress

- as close as practicable to potential ingress points, e.g. wall/floor joints, service inlet points, etc.

17.4 SUMMARY

Monitoring systems in buildings are generally based on the catalytic or infra-red type sensors, linked to automatic logging and warning systems. The positioning of the sensors should take into account buoyancy differences of the different gases present. Sensors should be sited close to likely gas ingress points, with additional sensors being placed at ceiling level.

Particular areas where monitoring may be necessary include rooms with little or no ventilation, rooms with potential ignition sources, underfloor and wall cavities, ducts, sewers and drainage channels.

References

1. INTERDEPARTMENTAL COMMITTEE ON THE REDEVELOPMENT OF CONTAMINATED LAND.
    Notes on the redevelopment and after-use of landfill sites.
    ICRCL Guidance Note 17/78, eighth edition, 1990.

2. DEPARTMENT OF THE ENVIRONMENT, HER MAJESTY'S INSPECTORATE OF POLLUTION.
    The control of landfill gas.
    Waste Management Paper No.27, HMSO, 1st Edition 1989, 2nd Edition 1991.

3. CREEDY, D.
    Geological Sources of Methane in Relation to Surface and Underground Hazards.
    Methane — Facing the problems symposium, Nottingham, UK, 26-28 September 1989.

4. DEPARTMENT OF THE ENVIRONMENT, PROPERTY SERVICES AGENCY.
    The Protection of Operational Buildings from Gas.
    PSA Report No.TAD(77)1, Directorate of Post Office Service, Croydon, 1977.

5. CREEDY, D.
    Hazardous Gases in the Mine Environment.
    Hazardous Gases in the Environment — A Seminar, North West Group, Institution of Geologists, UMIST, Manchester, UK, 10 April 1990.

6. GRAYSON, R. OLDHAM, L & CONNELL, S.
    Hydrogen Sulphide Springs in the Pennines and Adjacent Areas — A Neglected Hazard. *Hazardous Gases in the Environment — A Seminar*, North West Group, Institution of Geologists, UMIST, UK, 10 April 1990.

7. HOOKER, P.J. & BANNON, M.P.
    Methane: its occurrence and hazards in construction
    CIRIA Report 130, London, 1993.

8. BRITISH STANDARDS INSTITUTION.
    Draft for Development, Code of Practice for the Identification of Potentially Contaminated Land and its Investigation.
    DD 175:1988.

9. INTERNATIONAL STANDARDS ORGANISATION.
    Technical Committee 190.
    Soil Quality — Sampling, Part 3: Guidance on Safety.
    Secretary Committee, EPC48 "Soil Quality".
    British Standards Institution.

10. SMITH, M. A.
    Safety Aspects of Waste Disposal to Landfill.
    Proceedings of 'The Planning and Engineering of Landfills'
    Midland Geotechnical Society, Birmingham, July 1991.

11. HEALTH & SAFETY EXECUTIVE.
    Avoiding Danger from Underground Services.
    Her Majesty's Stationery Office, 1989.

12. HEALTH & SAFETY EXECUTIVE.
    Guidance Note GS5. Entry into confined spaces.
    Her Majesty's Stationery Office.

13. SYPOL ENVIRONMENT MANAGEMENT LTD.
    A guide to the control of substances hazardous to health in design and construction.
    Unpublished report to CIRIA Core Programme Funders FR/CP/4 June 1992 (to be published).

14. CONSTRUCTION INDUSTRY RESEARCH AND INFORMATION ASSOCIATION.
Safe Working Practices for Contaminated Sites.
CIRIA Report, London.

15. BRITISH DRILLING ASSOCIATION
Guidance notes for the safe handling of landfills and contaminated sites.
BDA, Brentwood, 1992.

16. SKYSCAN
Publicity information, Stanway Grounds Farm, Stanway, Cheltenham, Glos.

17. WELTMAN, A.
The use of Aerial Infra-red Photography for the Detection of Methane from Landfills.
Ground Engineering, Vol.6, No.3, pp22−23 April 1983.

18. BUTTERWORTH, J.S.
Sampling, Analysis and Identification.
Methane − Facing the Problems Symposium, Nottingham, UK, 26-28 September 1989.

19. BARROW, N. PEARSON, C.F.C. POWELL, P. EDWARDS, J.
The Source and Extent of Hydrocarbon Gas Occurrances near to an Underground Zinc Mine at Navan, Co Meath, Ireland.
Methane − Facing the Problems Symposium, Nottingham, UK, 26-28 March 1991.

20. FIRTH, J.G.
Measurement of flammable gases and vapours.
In: *Detection and Measurement of Hazardous Gases*. Edited by C.F. Cullis and J.G. Firth, Heinemann. London 1981.

21. PARTRIDGE, R.H. & CURTIS, I.H.
Final Report on the Methane Measurements made with the NPL Diode Laser Remote Monitoring System at New Park Landfill Site, Ugley, Stansted, Essex.
National Physical Laboratory Report Qu 526. November 1986.

22. BRITISH STANDARDS INSTITUTION.
Instruments for the Detection of Combustible Gases, Part 1-5.
BS6020 1981.

23. BRITISH STANDARDS INSTITUTION.
Gas Detector Tubes.
BS5343 1976.

24. COOPER, L.R.
Oxygen Deficiency, section 3.
In: *Detection and Measurement of Hazardous Gases*. Edited by C.F. Cullis and J.G. Firth, Heinemann. London 1981.

25. THORBURN, S. COLENUTT, B.A. AND DOUGLAS, S.G.
The sampling and gas chromatographic analysis of gases from landfill sites.
International Journal of Environmental Analytical Chemistry, 6 (3) 245 − 254, 1979.

26. FLOWER, F.B.
Case history of landfill gas movement through soils.
In report by Gentelli *et al. Gas and leachate from landfills, formation, collection and treatment.* Rep. No. 600-9-76-004 p177 - 189. US EPA March 1984.

27. EMCON ASSOCIATES.
Methane generation and recovery from landfills.
Consolidated Concrete Ltd and Alberta Environment - Ann. Arbor. Science. Michigan, 1980.

28. CONESTOGA-ROVER & ASSOCIATES.
Gas recovery & utilisation from a municipal waste disposal site.
Environmental Protection Service Report. No EPS − 4 − EC − 81 − 2. 1981.
Environmental Protection Service, Ottawa, Ontario, Canada.

29. WESTBAY INSTRUMENTS LTD, VANCOUVER, CANADA.
The MP system for water pressure measurements, groundwater samples and permeability tests. Publicity material.

30. BLAKEY, N.C. & MARIS, P.J.
Gas from Domestic Wastes in Landfills: Sampling and Analysis.
Water Research Centre. Stevenage Laboratory. LR1144. March 1980.

31. MCBEAN, E.A. FARQUHAR, G.J.
An examination of temporal and spatial variation in landfill generated methane gas.
Water air soil pollution, 13, p157 - 172, 1980.

32. Lytwynshyn, G.R. Zimmerman, R.E. Flynn, N.W. Wingender, R. Oliveri, V.
Landfill methane recovery: Part II — Gas characterisation.
Argonne National Laboratory, ANL/CNSV-TM-118. IL (USA), December 1982.

33. LEACH, A. & MOSS, H.D.T.
Landfill gas research and development studies: Calvert and Stewartby landfill sites, contractor report for Department of Energy, 1989.

34. JOHNSON, R.L. PANKOW, J.F. CHERRY, J.A.
Design of a Groundwater Sampler for Collecting Volatile Organics and Dissolved Gases in Small Diameter Wells.
Groundwater Vol.25, No.4, July — August 1987.

35. BARCELONA, M.J. HELFRICH, J.A. GARSKE, E.E. GIBB, J.P.
A Laboratory Evaluation of Groundwater Sampling Mechanisms.
Groundwater Monitoring Review. Vol. 4, p 32. Spring 1984.

36. NEWTON, J.
Groundwater Investigation and Monitoring.
Pollution Engineering. pp66-73, July 1989.

37. *METHODS FOR THE EXAMINATION OF WATERS AND ASSOCIATED MATERIALS.*
The Determination of Methane and Other Hydrocarbon Gases in Water 1988.
Her Majesty's Stationery Office, London, 1988.

38. BARBER, C. & BRIEGEL, D.
A Method for the Determination of Dissolved Methane in Groundwater in Shallow Aquifers.
Journal of Contaminant Hydrology, 2 (1987) pp51-60.

39. DEPARTMENT OF THE ENVIRONMENT.
The Householder's Guide to Radon.
HMSO, July 1990.

Appendix A Standards and codes of practice relevant to safety in site investigations for gas

BS 4683:	*Electrical Apparatus for Explosive Atmospheres:*
Part 1:1971	*Classification of maximum surface temperatures.*
Part 2: 1971	*The construction and testing of flameproof enclosures of electrical apparatus.*
Part 3: 1972	*Type of protection N.*
Part 4: 1973	*Type of protection 'e'.*
BS 5343:	
Part 2: 1991	*Specification for long-term gas detector tubes.*
BS 5345:	*Selection, installation and maintenance of electrical apparatus for use in explosive atmospheres (other than mining applications or explosive processing and manufacture).*
Part 1: 1976	*Basic requirements for all parts of the code.*
Part 2: 1983	*Classification of hazardous areas.*
Part 3: 1979	*Installation and maintenance requirements for electrical apparatus with type of protection 'd'. Flameproof enclosure.*
Part 4: 1977	*Installation and maintenance requirements for electrical apparatus with type of protection 'i'. Intrinsically safe electrical apparatus and systems.*
Part 6: 1978	*Installation and maintenance requirements for electrical apparatus with type of protection 'e'. Increased safety.*
Part 7: 1979	*Installation and maintenance requirements for electrical apparatus with type of protection 'N'.*
BS 5573: 1978	*Code of practice for safety precautions in the construction of large diameter boreholes for piling and other purposes.*
BS 5930: 1981	*Code of practice for site investigations.*
BS 6020:	*Instruments for the detection of combustible gases.*
Part 1: 1981	*Specification for general requirements and test methods.*
Part 2: 1981	*Specification for safety and performance requirements for Group I instruments reading up to 5% methane in air.*
Part 3: 1982	*Specification for safety and performance requirements for Group I instruments reading up to 100% methane.*
Part 4: 1981	*Specification for performance requirements for Group II instruments reading up to 100% lower explosive limit.*
Part 5: 1982	*Specification for performance requirements for Group II instruments reading up to 100% gas.*
BS 6164: 1990	*Safety in tunnelling in the construction industry.*
DD 175: 1988	*Code of practice for the identification of potentially contaminated land and its investigation.*

Health & Safety Executive. *Protection of personnel and the general public during development of contaminated land,* HS(G) 66, HMSO, 1991

Health & Safety Commission. *Control of substances hazardous to health* (General ACoP — second edition) and *Control of carcinogenic substances* (Carcinogens ACoP — second edition): Approved codes of practice'. HMSO, London, 1990.

Health & Safety Executive. *Avoiding danger from underground services.* HMSO, London, 1989.

Health & Safety Executive. Guidance Note GS5. *Entry into Confined Spaces.* HMSO, London, 1977.

International Standards Organisation. Technical Committee 190, *Soil quality.* For information contact Secretary Committee EPC48 'Soil quality,' British Standards Institution, London.

CIRIA Report 131

The measurement of methane and other gases from the ground

D Crowhurst and S J Manchester
Fire Research Station
Building Research Establishment

BUILDING RESEARCH ESTABLISHMENT, GARSTON, WATFORD WD2 7JR
Tel: 0923 894040   Telex: 923220   Fax: 0923 664010

CONSTRUCTION INDUSTRY RESEARCH AND INFORMATION ASSOCIATION
6 Storey's Gate, Westminster, London SW1P 3AU
Tel 071-222 8891   Fax 071-222 1708

Summary

Methane and the gases often found with it can cause many problems during construction operations such as tunnelling and building in the vicinity of landfill sites. At whatever stage of the construction process, from initial investigation to the operation of the completed works, if there is reason to suspect the presence of these gases, it will be necessary to be sure of safety. This report describes the methods and techniques that are available for the detection of these gases, for sampling them and for taking measurements relevant to mitigating the problems the gases pose.

While centred on methane and landfill gas, the report also covers carbon dioxide and other hazardous gases. Within the context of safe working, there is comprehensive guidance on detecting gas, identifying the source, measuring and sampling different gases, and on the interpretation of the results. Factors which affect the investigation of gas, e.g. meterological conditions, gas in groundwater, are discussed. The use of monitoring systems on a site and within buildings is described. In addition to the cited references, an Appendix lists standards and codes relevant to safety in site investigations for gas.

CROWHURST, D and MANCHESTER, S J
The measurement of methane and other gases from the ground
Construction Industry Research and Information Association
CIRIA Report 131, 1993

Keywords:
Methane, gases, measurement

Reader interest:
All waste disposal and environmental professionals, developers

ISBN 0 86017 3720

ISSN 0305 408X

© CIRIA 1993

	CLASSIFICATION
AVAILABILITY	Unrestricted
CONTENT	Report
STATUS	Committee Guided
USER	Construction professionals

Published by Construction Industry Research and Information Association, 6 Storey's Gate, Westminster, London SW1P 3AU in association with Building Research Establishment, Garston, Watford, Herts. WD2 7JR.